Qualitative Research Methods in Human Geography

Other Meridian Titles

Howard Bridgman, Robin Warner, and John Dodson,
Urban Biophysical Environments
David Chapman, *Natural Hazards*, Second Edition
Arthur and Jeanette Conacher, *Rural Land Degradation in Australia*
Robert H. Fagan and Michael Webber, *Global Restructuring: The Australian Experience*, Second Edition
Dean Forbes, *Asian Metropolis: Urbanisation and the South-East Asian City*
Clive Forster, *Australian Cities: Continuity and Change*, Second Edition
Nick Harvey, *Environmental Impact Assessment: Procedures, Practice, and Prospects in Australia*
Iain Hay, *Communicating in Geography and the Environmental Sciences*
Jamie Kirkpatrick, *A Continent Transformed: Human Impact on the Natural Vegetation of Australia*, Second Edition
Elaine Stratford, *Australian Cultural Geographies*

Qualitative Research Methods in Human Geography

Edited by
Iain Hay

Series editors
Deirdre Dragovich
Alaric Maude

OXFORD
UNIVERSITY PRESS
253 Normanby Road, South Melbourne, Victoria, Australia
Oxford University Press is a department of the University of Oxford.
It furthers the University's objective of excellence in research, scholarship,
and education by publishing worldwide in
Oxford New York
Athens Auckland Bangkok Bogotá Buenos Aires Calcutta
Cape Town Chennai Dar es Salaam Delhi Florence Hong Kong Istanbul
Karachi Kuala Lumpur Madrid Melbourne Mexico City Mumbai Nairobi
Paris Port Moresby São Paulo Shanghai Singapore Taipei Tokyo Toronto Warsaw
with associated companies in Berlin Ibadan

First published 2000

National Library of Australia
Cataloguing-in-publication data:

Qualitative research methods in human geography.

Includes index.
ISBN 0 19 550787 8.

1. Human geography—Research—Methodology.
I. Hay, Iain, 1960–. II. Title. (Series: Meridian
series in geography).

304.2

Edited by Felicity Rawlings
Indexed by Russell Brooks
Cover designed by Modern Art Production Group
Typeset by McMillan Design, Melbourne
Printed through Bookpac Production Services, Singapore

Foreword

Meridian: Australian Geographical Perspectives is a series initiated by the Institute of Australian Geographers in 1990 to meet the need for relatively short, low-cost books written for university students. The books in this series are designed to explore the geographical issues and problems of Australia and its region, to present an Australian perspective on global geographical processes, or to provide students with Australian-oriented texts on geographical skills. The term 'meridian' refers to a line of longitude linking points in a half-circle between the poles. In this series it symbolises the interconnections between places in the global environment and global economy, which is one of the key themes of contemporary geography. The books in the series cover a variety of physical, environmental, economic, and social geography topics, and are written for use in first- and second-year courses where the existing texts and reference books lack a significant Australian perspective. To cope with the varied content of geography courses taught in Australian universities, the books are not designed as comprehensive texts, but as modules on specific themes that can be used in a variety of courses. They are intended for use in either a one-semester course or a one-semester component of a full-year course.

Titles in the series cover a range of topics representing contemporary Australian teaching and research in geography. These topics include economic restructuring, vegetation change, land degradation, cities, cultural geography, feminist geography, natural hazards, and urban environmental problems, as well as presentation skills. Future topics include the economic geography of Australian cities and coastal management. Although the emphasis in the series is on Australia, we will also have occasional titles on South-East Asia, using the considerable expertise that Australian geographers have developed on this region.

While the primary aim of the series is to produce books for students, the selected topics deal with issues of relevance to all Australians. We therefore hope that the general reader will find some of these titles of interest, and discover that geographers have something distinctive to say about contemporary environmental, economic, and social issues. As the books assume little or no previous training in geography, and attempt to avoid a textbook style, they should be readily understood by the general reader.

This book is the twelfth in the series, and introduces the reader to a range of qualitative methods used in human geographic research. The book begins with a discussion of the place of qualitative research in human geography, and then examines issues of ethics and research design. The book then explains how to use interviews, focus groups, observation, and textual analysis in research. These chapters are full of useful advice, and will enable beginning researchers to develop sound and defendable methods. The final chapters explore the use of computers in qualitative data analysis, and some issues confronting geographers communicating research findings. A unique characteristic of the book is the way it combines practical advice on how to use particular techniques with an introduction to the difficult issues of rigour, objectivity, and truth.

Iain Hay, the editor, has been using qualitative techniques for over two decades, and teaching them since 1992. He has also gained an international reputation for his work on learning and teaching in geography. He has a graduate qualification in tertiary education, a Flinders University award for excellence in teaching, and a strong commitment to good teaching. His earlier book in the Meridian Series, *Communicating in Geography and the Environmental Sciences*, has been very popular. For this book he has assembled a group of Australian and New Zealand geographers, so that each chapter is written by an expert or experts in the field. The editor has welded these chapters together through the use of standard elements such as lists of key terms and review questions, as well as by extensive editing. The result is a readable, practical, and thoughtful guide to qualitative research methods that will improve the work of beginning and experienced researchers alike.

Deirdre Dragovich
University of Sydney

Alaric Maude
Flinders University

For Tania
and for my Scottish relatives,
especially Caroline and Eleanor

Contents

List of Boxes

Notes on Contributors

Lawrence Berg, BA (Dist) (Victoria) 1988, MA (Victoria) 1991, PhD (Waikato) 1996, was Lecturer in the School of Global Studies at Massey University from 1995 to 2000. He is currently College Professor at Okanagan University College and an Adjunct Professor in the Department of Geography, University of Victoria. He has a diverse range of academic interests in radical and critical geography. Lawrence has published in a wide variety of books and journals, including *Canadian Geographer*, *Environment and Planning A*, *Environment and Planning D*, *Gender, Place and Culture*, *Journal of Geography in Higher Education*, *Historical Geographer*, *New Zealand Geographer* and *Progress in Human Geography*.

Matt Bradshaw, BA (Hons) (Tasmania) 1989, MEnvStudies (Tasmania) 1992, PhD (Tasmania) 2000, is a Research Fellow in the Department of Geography and Environmental Studies, University of Tasmania. He has taught at the University of Tasmania. He is also a social research consultant to the Hobart City Council and the Tasmanian Aquaculture and Fisheries Institute, among others. His current academic interests include economic geography, the social impact assessment of fisheries, and community involvement in local government planning. He has published in *Australian Geographical Studies*, *Geography and Parks* and *Leisure Australia*.

Jenny Cameron, DipTeach (Brisbane CAE) 1983, BAppSc (QUT) 1990, MA (Sydney) 1992, PhD (Monash) 1998, is a Research Fellow in the School of Public Policy at Monash University working on an action research project with groups of retrenched workers, single parents and unemployed young people. Before completing her PhD at Monash University in 1998, she had a variety of teaching and research positions at Monash University, Queensland University of Technology, Queensland Department of Housing, Local Government and Planning, and Ian Buchan Fell Housing Research Centre and has also taught at primary schools in Brisbane and Thursday Island. Her academic interests include domestic labour, regional development and contemporary social theory. She has published papers and chapters in *Rethinking Marxism; Australian Geographer; Class and Its Others* (Minnesota University Press 2000); *Australian Feminism: A Companion* (OUP 1998, with K. Gibson); and *Reader's Guide to Women's Studies* (Fitzroy Dearborn 1998).

Robyn Dowling, BEc (Hons) (Sydney) 1988, MA (British Columbia) 1991, PhD (British Columbia) 1995, is Lecturer in the School of Human Geography at Macquarie University. Prior to that she was teaching and completing her PhD at the University of British Columbia. Her current research explores the cultural geographies of Australian cities; gender and suburbia; and transport and mobility. Robyn's publications include *Social Atlas of Sydney* (Sydney University Press 1989, with R. Horvath and G. Harrison), a chapter in *Cities of Difference* (Guilford 1998) and papers in *Antipode, Australian Geographical Studies, Canadian Geographer, Housing Studies* and *Urban Geography*.

Kevin Dunn, BA (Hons) (Wollongong) 1990, PhD (Newcastle) 1999, is Lecturer in the School of Geography at the University of New South Wales. Like other contributors to this volume, Kevin has a broad range of academic interests, including planning for ethnic diversity; theories of ethnic concentration and migrant settlement; land use conflict; reconstruction of industrial cities; representation of place; and the politics of heritage and memorial landscapes. His PhD focused on opposition to non-Christian places of worship in Sydney. He co-authored *Introducing Human Geography* and has published more than a dozen chapters and articles in various books and in journals including *Australian Geographer, Australian Geographical Studies, Environment & Planning* A and *Urban Studies*. He also edited the 1997 special collection on 'Cultural geography' for *Australian Geographical Studies*.

Dean Forbes, BA (Flinders) 1970, MA (Papua New Guinea) 1976, PhD (Monash) 1979, is Professor in the School of Geography, Population and Environmental Management and Pro-Vice-Chancellor (International) at Flinders University. He is also a Fellow of the Academy of Social Sciences in Australia. He has held research and teaching appointments in the Research School of Pacific Studies of the Australian National University, at Monash University, and at the University of Papua New Guinea. His research interests include postcolonialism and the new geographies of Pacific Asia, the social, economic and environmental sustainability of Asian megacities, and Australia's integration with Asia. Some of his recent publications include *Asian Metropolis: Urbanisation and the Southeast Asian City* (OUP 1996), *Multiculturalism, Difference and Postmodernism* (Longman Cheshire 1993 [co-editor]) and *Urbanisation in Asia* (AusAID 1997).

Iain Hay, BSc (Hons) (Canterbury) 1982, MA (Massey) 1985, PhD (Washington) 1989, GradCertTertEd (Flinders) 1995, is Professor in the School of Geography, Population and Environmental Management at Flinders University. He is author of *The Caring Commodity* (OUP 1989), *Money, Medicine and Malpractice in American Society* (Praeger 1992), *Communicating in Geography and Environmental Studies* (OUP 1996) and *Making the Grade* (OUP 1997, with D. Bochner and C. Dungey). Iain is also Asia-Pacific editor of *Ethics, Place and Environment*, and Australasian editor of *Journal of Geography in Higher Education*. His research interests include geographies of justice and regulation. He has a particular interest in the role of insurers as regulators. He also conducts work on geographical education. Iain is actively involved in professional activities with the Institute of Australian Geographers and the Royal Geographical Society of South Australia.

Robin Kearns, BA (Auckland) 1981, MA (Hons) (Auckland) 1983, PhD (McMaster) 1988, is Senior Lecturer at the University of Auckland. His research interests include relationships between health and place; cultural geographies of place; and the ethics and relations of research. He has edited *Putting Health into Place: Landscape, Identity and Well-being* (Syracuse University Press, with W. M. Gesler) and has written more than fifty articles in journals including *Environment and Planning D: Society and Space, Health and Place, Housing Studies, Professional Geographer* and *Social Science and Medicine*.

Juliana Mansvelt, BA (Hons) (Massey) 1989, PhD (Sheffield) 1994, is currently employed as a Lecturer in Geography in the School of Global Studies, Massey University. Her teaching and research interest include the political economy of local economic development, geographies of leisure and of consumption processes and spaces. Her recent research on the social construction of ageing has utilised qualitative interviews to explore relationships between ageing, place and leisure. Juliana has published papers and chapters in *Area*, *New Zealand Journal of Geography*, *New Zealand Geographer*, *Physical Education New Zealand*, *Time Out? Leisure Recreation and Tourism in New Zealand and Australia* (Longman 1998) and the recent Oxford University Press text *Encountering Place*.

Robin Peace, DipTchg (Christchurch College of Education) 1978, BA (Canterbury) 1993, BSocSci (Hons) (Waikato) 1993, DPhil (Waikato) 1999, is Lecturer in the Department of Geography at the University of Waikato. She has had a broad range of research and teaching experience, having been Head of Geography at Wellington East Girls' College, Lecturer in Professional Studies at the Hutt Valley Outpost of the Christchurch College of Education, and visiting research worker at the University of Edinburgh. Robin has also co-edited the *New Zealand Geographer* and has been a member of the UNESCO New Zealand Social Sciences SubCommission since 1995. Her research interests include feminist and postmodern epistemologies and contemporary geographical thought, social exclusion, citizenship and corporeality, geography and social policy (with a focus on New Zealand and the European Union), geographical education at the tertiary and secondary levels, and qualitative methods and computer assisted qualitative data analysis. She has published on these topics in a range of publications including *Gender, Place and Culture*, *New Zealand Journal of Geography* and *New Zealand Geographer*.

Elaine Stratford, BA (Flinders) 1984, BA (Hons) (Flinders) 1986, PhD (Adelaide) 1996, is Lecturer in the Department of Geography and Environmental Studies, University of Tasmania. She has taught previously at the Australian Defence Forces Academy and Flinders University. Her current academic interests include urban planning and development; place making; skate cultures and inner-city local government; and sustainable communities research. She has published widely in journals including *Australian Geographical Studies*, *Australian Journal of Communication*, *Gender, Place and Culture*, *Health and Place*, *Sites*, *Social Alternatives* and *South Australian Geographical Studies*. Elaine recently edited *Australian Cultural Geographies* for Oxford University Press' Meridian Series, and has chapters published in *The Child's World: Triggers* *to Learning* (Robertson and Gerber, eds, ACER) and in *Small Worlds, Global Lives: Islands and Migration* (Connell and Russell, eds, Pinter).

Hilary Winchester, BA (Hons) (Oxford) 1974, MA (Oxford) 1977, DPhil (Oxford) 1980 is currently Professor and President of Academic Senate at the University of Newcastle. She has taught previously at the Universities of Wollongong and Plymouth. She edited *Journal of Interdisciplinary Gender Studies* from 1996 to 1997 and is currently co-editor of *Australian Geographical Studies*. Hilary has a wide range of interests in population, social and cultural geography, specifically including marginal social groups (especially one-parent families), the social construction of place, the geography of gender, and urban social and landscape planning. She is author of *Contemporary France* (Longman 1993) and approximately fifty articles and book chapters, including publications in *Australian Geographer, Australian Geographical Studies, Environment and Planning D: Society and Space, Espace, Populations, Sociétés* and *Transactions of the Institute of British Geographers*.

Preface

Between five and ten times a year, he travels for a week or two to the world's most important markets and growth regions. There he makes inquiries about every conceivable aspect of economic life. Hardly any doors are closed to him; his conversation partners from industry, government and central banks know the inestimable value of such a pathfinder for the transnational flow of capital. In these conversations Trent is not looking for figures and mathematically grounded predictions. 'Everyone's got the latest data on computer,' he holds forth. 'What counts is people's mood, the conflicts beneath the surface' (Martin and Schumann 1997, p. 55).

This volume provides concise and accessible introductory materials on qualitative research methods in human geography. It gives particular emphasis to examples drawn from 'social/cultural geography', perhaps the most vibrant area of inquiry in Australian and New Zealand/ Aotearoa geography over the past decade. The book is aimed primarily at an audience of undergraduate students in second- and third- year topics, but I expect that a good deal of the material covered will also be of considerable interest to honours students and perhaps to people commencing postgraduate study. The chapters have been written with the dual intent of providing novice researchers with clear ideas on how they might go about conducting their own qualitative research thoroughly and successfully and of offering university academics a teaching-and-learning framework around which additional materials and exercises on research methods can be developed. The text is unique in its dedication to the provision of practical, 'how to' guidance on methods of qualitative research in geography.

Without knowing it at the time, I started work on this book in 1992 when I was asked to teach a topic called 'Research Methods in Geography' at Flinders University. I developed lectures and extensive sets of notes for my classes. I also referred students to texts such as Bernard (1988), Kellehear (1993), Patton (1990), Sarantakos (1993) and Sayer (1992). All of these books came from disciplines other than geography or referred to the social sciences in general. It disturbed me that despite the renewed emphasis on qualitative research and the teaching of qualitative methods in the discipline—reflected then and now and in work such as Dyck (1999), Harrison and Burgess (1994), Holland and Hargreaves (1991), Lane (1997), Lee (1992), McBride (1999), Nettlefold and Stratford (1999), Pile (1992) and Tuan (1991)—no geographer had produced an accessible text on the day-to-day practice of qualitative research.

In the meantime, during their regular visits to my office, publishers' representatives asked what sorts of books might be useful in my teaching. I repeatedly mentioned the need for a good text that dealt with qualitative research methods in geography. That message came back to haunt me when one of the Meridian Series Editors, Dr Alaric Maude, asked me if I would like to write a book on qualitative methods for the series. I declined, but offered instead to try to draw together the expertise and energies of a group of active and exciting geographers from Australia and Aotearoa/New Zealand to produce an edited collection.

Of course, in the time since I planned the text and the contributors and I fulfilled the terms of our publishing contract, a new group of research methods texts have emerged from geographers around the world! These include Flowerdew and Martin (1997), Kitchin and Tate (2000), Lindsay (1997) and Robinson (1998). Despite their various merits, none of these books deals explicitly with qualitative methods and none takes its roots in Australia and New Zealand/Aotearoa. This book does both.

In overview, *Qualitative Research Methods in Human Geography* locates qualitative research methods in disciplinary and social contexts and gives particular emphasis in its examples to the Asia-Pacific Region. Despite that regional focus, the examples should still inform the learning, teaching and research work of 'northern' readers. Indeed, I would hope that just as teachers in Australia, New Zealand/Aotearoa, Southeast Asia and the Pacific elaborate on British and North American texts by referring to local examples, colleagues in 'our' antipodes will be able to supplement this volume's examples with local ones of their own. The book provides detailed advice on the conduct of the most widely used forms of qualitative research employed in human geography: interviews, focus groups, observation and textual analysis. It outlines means of analysing qualitative information, including computer assisted data analysis. Finally, it deals with the problematic issue of 'writing-up' qualitative research. The nine chapters range over the material outlined in the following paragraph.

In Chapter 1, Hilary Winchester situates qualitative research methods within the context of geographic inquiry. She outlines the range of qualitative techniques used commonly in human geography and explores the relationship between those methods and the recent history of geographic thought. On this foundation, Robyn Dowling builds a review of some critical issues associated with qualitative research. These include power relationships between researchers and their co-researchers, questions of subjectivity and intersubjectivity, and points of ethical concern. Chapter 3 by Matt Bradshaw and Elaine Stratford examines the difficult but vitally important matters of design and rigour in qualitative research. The four subsequent chapters offer concise, similarly structured outlines of some of the main forms of qualitative research practice employed in geography. Kevin Dunn draws on his extensive experience of interviewing in cross-cultural settings to provide a comprehensive, yet concise, examination of interviewing practice. This flows into a chapter by Jenny Cameron on the utility and conduct of focus groups in geography. Robin Kearns, of the University of Auckland, reviews the purposes and practice of observational techniques in human geography. The final chapter in this section is by Dean Forbes. In a chapter richly supported with examples, Dean critically examines the problematic practice of 'reading' texts, including manuscripts, maps, photographs, paintings, film and multimedia images. The penultimate chapter, by Robin Peace, explores some of the ways in which computers may be used to make sense of qualitative data. Lawrence Berg and Juliana Mansvelt conclude the book with a discussion of some of the practical and conceptual issues surrounding representation of qualitative research findings.

As I have already observed, the following chapters offer a comprehensive overview of qualitative research methods in human geography. However, the book is not intended to be a 'one-stop' resource or prescriptive outline (indeed, how could it be?) for qualitative researchers and students. It is instead a starting point and framework. Accordingly, each chapter includes direction to a number of additional key readings and relevant sources that may be consulted to follow up material introduced here. There is, of course, a consolidated list of all references cited for those readers who might wish to enquire even further. Chapters also include review questions intended for classroom discussion or as individual exercises. They might also serve as prompts for quite different exercises. Indeed, if you have discovered or created any useful exercises to illustrate some of the matters covered within these chapters, I would be delighted to hear about them.

The book also features an extensive glossary. While the authors have made efforts to ensure that their chapters are written in a language that is accessible to undergraduate readers, there are—without doubt—terms that will be somewhat alien to many readers in the first instance. The glossary should help resolve that sort of difficulty. Terms included in the glossary are drawn primarily from the lists of key words associated with each chapter. I have included the glossary knowing full well how useful it is likely to be to student readers hurrying to refresh their memory about a term before losing the thread of a chapter's discussion.

I owe thanks to many people who have been involved with the production of this book. Students in my third year 'Research Methods in Geography' classes at Flinders University encouraged me to put the book together and offered honest and helpful comments on its chapters. Every chapter in this collection was reviewed by two experts in the field who, without exception, provided timely, critical, and comprehensive opinion. Those reviewers were: Kay Anderson (Durham), Andrew Beer (Flinders), Michael Brown (Washington), Jacquie Burgess (University College, London), Mike Crang (Durham), Jon Goss (Hawai'i), Andy Herod (Georgia), Richie Howitt (Macquarie), Jane Jacobs (Melbourne), Lucy Jarosz (Washington), Ron Johnston (Bristol), Murray McCaskill (Flinders), Pauline McGuirk (Newcastle), Eric Pawson (Canterbury), Chris Philo (Glasgow), Lyn Richards (La Trobe), Pamela Shurmer-Smith (Portsmouth) and Lynn Staeheli (Colorado). The Meridian Series Editors, Deirdre Dragovich and Alaric Maude, and Jill Henry of

Oxford University Press have also provided thoughtful comment and ongoing support for which I am very grateful.

I must also express a special note of thanks to the contributors to this volume. They put up with repeated emailed requests and comments from me that were often as long as the chapters themselves! I remember sending out those messages uttering, as I pressed the 'send' button, a quiet prayer to the effect: 'please, please don't let them spit the dummy and pull out of this project!' I admire the authors' tolerance, persistence and fortitude in the face of both my editing style and the various other forms of personal and professional adversity they encountered during the course of this project.

Finally, I owe very special thanks to Cecile Cutler. Cecile has the misfortune to occupy the office next to mine. As a result, she not only has to contend with door-slamming (it's the wind, really...), but also with my incessant requests to look over work for sense and style. Her eye is ever-critical, her humour infectious and her comments frank. Thanks Cecile!

Iain Hay

1

Qualitative Research and its Place in Human Geography

Hilary P. M. Winchester

CONTENTS

CHAPTER OVERVIEW

This chapter aims first to provide an overview of contemporary qualitative research methods in human geography. The range of methods commonly used in human geography is considered and categorised, together with some of the ways in which those methods are used to provide explanation. Second, the chapter aims to review briefly the context from which qualitative research has developed in human geography. This is achieved by examining changing schools of thought within the discipline, recognising changes are messy, overlapping and coexisting. Third, the chapter aims to link methodological debates to wider theoretical perspectives in geography.

INTRODUCTION

Contemporary human geographers study places, people, bodies, **discourses**, silenced voices and fragmented **landscapes**. The research questions of today's human geographers require a multiplicity of conceptual approaches and methods of enquiry. Increasingly, the research methods used are qualitative ones intended to elucidate human environments, individual experiences and social processes. This introductory chapter aims to set the scene of qualitative research in human geography and to highlight issues and techniques that are examined in more detail in later chapters. The chapter has three main and interlinked objectives.

First, it provides an overview of qualitative methods, their context, and the links between methodology and theory. The categories used and established here are relatively fluid; they are designed merely as a way of organising this growing field, and are not meant to be fixed or constraining. Indeed, some recent research, such as that on the body and embodied experiences in place, essentially defies categorisation. A number of issues raised in this introduction are more fully developed later in the volume, particularly those that relate to ethical practices (Chapter 2), the positioning of the author relative to the audience (Chapters 2, 7 and 9), as well as the broad issues of credibility, dependability and confirmability (Chapter 3).

Second, the chapter outlines the arguments about, and differences (and some of the similarities) between, qualitative and **quantitative methods**. Furthermore, the current debate on combining methods through **triangulation** and mixed-method approaches is reviewed. The resurgence of qualitative techniques and the current qualitative/quantitative debate in human geography are set in the context of the discipline's development and evolution. Inevitably a thumb-nail sketch of the evolution of geographical thought in part of just one chapter will necessitate some broad generalisations. However, in this context it is important to recognise that controversy about the nature and validity of research methods has existed for decades. Methodological debates may entrench polarised positions, for example between quantitative and qualitative, **objective** and **subjective**. The apparent polarity between those positions may in fact prove to be largely a false dichotomy. In the past, such debates in geography have raged around other extremes (Wrigley 1970). Geographers in the early decades of the twentieth century have argued, for example, over the merits of determinism versus possibilism (the extent to which humans have control over or are controlled by their environment) (Wrigley 1970). The current airing of the qualitative versus quantitative debate may in one sense be likened to the resurrection of a dinosaur in the shape of a false dichotomy, while in another sense it is a sign of healthy debate and intellectual vigour within the discipline.

Finally, the chapter focuses particularly on links between theory and methodology and raises issues of ethics, authorship and power. The contemporary use of qualitative methods in human geography is positioned within the theoretical debates and intellectual evolution of the discipline, while raising issues for consideration by reflexive and committed researchers. Intense arguments about methodology are often as much to do with researchers' beliefs and feelings about the structure of the world as about their regard for a particular research method, such as **participant observation** or in-depth **interviews**. While creationists and catastrophists clearly subscribe to a particular world-view (giving credence to the biblical account of earth's creation and catastrophic geomorphological events such as the Great Flood), other geographers may hold equally strong, but perhaps less obvious, views about the order, structure, measurement, and knowability of phenomena. In complex ways, **ontology** (beliefs about the world) and **epistemology** (ways of knowing the world) are linked to the methods we choose to use for research (Sayer and Morgan 1985).

WHAT IS QUALITATIVE RESEARCH?

What questions does qualitative research answer?

Qualitative research is used in many areas of human geography. In a broad sense, qualitative research is concerned with elucidating human environments and human experiences within a variety of conceptual frameworks. The term 'research' is used here to mean the whole process from defining a question to analysis and interpretation. 'Method' is used as a much more specific term for the investigative technique employed. A huge range of methods is used in many different situations. Some of the variety that springs to mind range from interviews about health issues in **postcolonial** New Zealand/Aotearoa (Dyck and Kearns 1995); through participant observation with homeless children in abandoned docklands of the city of Newcastle, New South Wales (Winchester and Costello 1995); to deconstruction of films and **textual** material analysing Australian gender-environment relations (Aitken and Zonn 1993). Inevitably, it is difficult to summarise the questions addressed by such a variety of research. However, it is instructive to recall the answer to a similar question posed two decades ago in relation to statistical analysis in geography. In that text, Ron Johnston (1978, pp. 1–5) argued that the gamut of statistical techniques answered two fundamental questions. Those questions were either about the *relationships* between phenomena and places, or the *differences* between them. The elegant simplicity of Johnston's questions can be paralleled by a different, but similarly broad, pair of questions which qualitative research is trying to answer.

The two fundamental questions tackled by qualitative researchers are concerned either with **social structures** or with individual experiences. This dualism is one which in practice may be hard to disentangle, but is of fundamental importance in explanation. The behaviour and experiences of an individual may be determined not so much by their personal characteristics but by their position in the social structure, 'together with their associated resources, constraints or rules' (Sayer 1992, p. 93). The first question may be phrased as:

Question One: What is the shape of societal structures and by what processes are they constructed, maintained, legitimised and resisted?

The structures which geographers are analysing may be social, cultural, economic, political or environmental. Structures may be defined as internally related objects or practices. Andrew Sayer (1992, pp. 92–5) gives the example of the landlord-tenant relationship, where structures exist in relation to private property and ownership, where rent is paid between the two parties, and where the structure may survive a continual turnover of individuals. Furthermore, he emphasises that tenants almost certainly exist within other structures; for example they may be students affected by educational structures, or migrants constrained by racist structures. The coexistence of rented housing, students and minority groups produces a complex linkage and mutual reinforcement of structures, within which individuals live out their lives. Qualitative geographers balance a fine line between the examination of structures and processes on the one hand and of individuals and their experiences on the other. Structures constrain individuals and enable certain behaviours, but in some circumstances individuals also have the capacity to break rather than reproduce the mould. An overemphasis on structures and processes rather than individuals could lead to a dehumanised human geography. On the other hand, individuals do not have all-powerful free will which would enable them to overcome the powerful structures embedded in society, such as capitalism, patriarchy or racism.

Sayer (1992, p. 95) considers that the key question for qualitative researchers about structures may be phrased as: 'What is it about the structures which produce the effects at issue?' Geographers have studied structures qualitatively in a number of ways. A significant focus has been on the ways in which they are built, reproduced and reified: for example Kay Anderson (1993) has analysed the documentary history which has led to stigmatisation of the suburb of Redfern in Sydney. In this example the structures are essentially indistinguishable from the processes which build, reinforce and contest them. Other authors have considered either the material or symbolic representations of structures: in their 1994 analysis of British merchant banking, McDowell and Court emphasise media representations of banking patriarchs and the importance of dress and body image for younger female and male bankers. A further aspect of the investigation of structures may be concerned with their oppressive or exclusionary nature: Gill Valentine's (1993) interviews with lesbian workers considered the ways in which workplace structures naturalise heterosexual norms, and thereby contribute to the oppression and marginalisation of workers who do not conform to these norms. Most of the qualitative geographical work on structures in fact emphasises the processes and relations which sustain, modify or oppose those structures, rather than focussing specifically on their form and nature.

The second question is concerned with individual experiences of structures and places:

Question Two: What are individuals' experiences of places and events?

Individuals experience the same events and places differently. Giving voice to individuals allows viewpoints to be heard which otherwise might be silenced or excluded: Jane Jacobs' (1993) account of the conflicts over mining at Coronation Hill gives voice to Aboriginal perspectives on naturalised notions of land and country. Dissident or marginalised stories may be 'given voice' through the use of diaries, oral histories, recordings of interviews and conversations or through the use of 'alternative' rather than 'mainstream' media. Participant observation by immersion in particular settings allows multiple viewpoints to be heard and acknowledged. A study of the post-school celebrations of Schoolies' Week on the Gold Coast of Australia (Winchester et al. 1999) gives voice not only to the partying Schoolies (i.e., high school students celebrating the end of the academic year), but also to agents of control (for example, police). The experiences of individuals and the meanings of events and places cannot necessarily be generalised, but they do constitute part of a multi-faceted and fluid reality. Qualitative geographical research tends to emphasise multiple meanings and interpretations rather than seeking to impose any one 'dominant' or 'correct' interpretation.

The experiences of an individual have been used in a generalisable sense to illuminate structures and structural change; Rimmer and Davenport (1998) use the travel diaries of Australian geographer Peter Scott as an example of the huge changes in mobility and technology which have characterised air transport since the 1950s. An autobiographical example by Reginald Golledge (1997) tells his personal story to outline the difficulties experienced by geographers and other academics faced with major physical disabilities. In some cases, the boundary between structure and individual experience is blurred: in a study of the Anglophone world in Quebec, Rose (1988) felt that the experience of being English in Quebec was intrinsically bound up with the structures and behaviours of that society in that place. In trying to express that intimate relationship, Rose (1988, p. 174) commented that there was '*no basis on which to express any difference* (author's emphasis) between the "tradition-as-lived" (i.e., being English in Quebec) and "a framework as an analysis of that tradition-as-lived"'. In other words, the experience itself could not be analytically separated from the structures which form the context for that experience.

Types of qualitative research

It is clear, even from the brief preceding section, that qualitative research in geography is currently used to address a huge range of issues, events and places, and that these studies utilise a variety of methods. Nonetheless, some methods are much more commonly employed than others. This section identifies three main types of qualitative research: the oral (primarily interview-based), the textual (creative, documentary and landscape) and the observational.

Clearly the most popular and widely used methods are oral. Talking with people as research subjects encompasses a wide range of activities. The spoken testimony of people is used in ways that range from the highly individualistic (oral histories and autobiographies) to the highly generalised (the individual as one of a random sample). The latter type of survey technique borders on the quantitative, where responses can be counted, cross-tabulated and analysed statistically (for example, Oakley 1981). The former approach, often achieved through oral history methods, lies at the more qualitative, individualistic end of the spectrum. Such methods are considered in more detail in Chapter 4. A middle ground is occupied here by the increasingly popular technique of using focus groups (see Chapter 5). Jackson and Holbrook (1995) effectively used focus groups differentiated by age, gender, and social group to analyse the complex meanings of the 'everyday' activity of shopping. The range of ways in which **oral methods** are utilised in geography—whereby subjects are allowed to speak with their own voice—is outlined in Box 1.1. It should be noted that the research questions will to some extent shape the methods that will be used. In particular, the methods range from answering the research question about individual meanings and experiences at the biographical end of the spectrum, to answering the research question about societal structures at the survey end. Surveys are undertaken to obtain information from and about individuals that is not available from other sources. While an interview is undertaken with an individual, a survey involves a more standardised interaction with a number of people. Oral surveys of personal information, attitudes and behaviour usually (but not inevitably) utilise questionnaires. Questionnaires are more closely structured and ordered than interviews, and every respondent answers the same question in a standard format. Mass mail-out questionnaires which do not involve personal contact and which are analysed quantitatively are not considered here as qualitative techniques, even though they may include some open-ended questions.

Box 1.1: Oral qualitative methods in human geography

General method	*Specific method*	*Research questions*
	Autobiography	Individual
Biography	Biography	↑
	Oral history	
	Unstructured	
Interviews	Semi-structured	
	Structured	
	Focus groups—open ended	
Surveys	Surveys—structured	↓
	Questionnaires—structured	General / Structural

The second major type of qualitative research is textual analysis (discussed more fully in Chapter 7). Such texts are wide-ranging but more diffuse in the human geography literature than the oral testimonies described above. Important groups of textual methods utilise creative, documentary and landscape sources. Creative texts are likely to include poems, fiction, films, art and music. Documentary sources may include maps, newspapers, planning documents, and even postage stamps! Landscape sources may range from the very specific, such as the 'cemeteries and columbaria, memorials and mausoleums' that Lily Kong (1999) uses in her analysis of deathscapes, to the general, such as the landscapes of suburbia as indicators of their social status (Duncan 1992). The analysis of creative sources, including fictional literature, film, art and music, has shown increasing complexity in recent years. (For an up-to-date survey of this field see Winchester and Dunn 1999. See also Chapter 7 of this volume.) Geographers have searched such sources for underlying structures, looking at paintings, for example, to understand changing perceptions of landscape (Heathcote 1975, pp. 214–17; Lowenthal and Prince 1965) or using film to examine both the impact of city restructuring and the ways in which it is represented (Winchester and Dunn 1999).

Written texts have also been used as the source of underlying discourses which underpin and legitimate social structures. Analysis of media representations demonstrate the myth-making power of repetitive sloganising, whether this relates to myths of the inner city (Burgess and Wood 1988) or to the imagery of place-making (Dunn et al. 1995). Herman (1999) has argued that changing place names in Hawai'i reveal a transformation from Hawai'ian political and cultural economy into western capitalist forms, while Sparke (1998) shows convincingly some of the ways in which *The Historical Atlas of Canada* is enmeshed in post-colonial politics of that country.

A significant and controversial source of textual analysis is the landscape itself (see Chapter 7 for a discussion of the debate). The argument that landscape may be read as text is epitomised in the work of Duncan and Duncan (1988) where the residential landscape is decoded of its social nuances. A recent study of roadside memorials in New South Wales concluded that the roadside crosses and flowers were indicative not only of individuals' behaviour but also of a 'problematic masculinity' characterised by aggression, fast driving and reckless behaviour (Hartig and Dunn 1998). Textual analyses of particular landscapes such as model housing estates use techniques derived from **semiotics** (the language of signs) to demonstrate literally the in-built naturalisation of social roles according to gender and family status (Mee 1994). Schein (1997) interprets landscape architecture, insurance mapping and other elements of a 'discourse materialised' to explore the ways they symbolise and constitute particular cultural ideals. Waitt and McGuirk (1996) examine both documentary and landscape texts to explore the selective representation of the heritage site of Millers Point in Sydney; the choice of particular buildings as 'heritage' both reflects and reproduces a white, male, coloniser's view of Sydney's history while silencing other views and voices.

The third significant type of qualitative research in human geography consists of forms of participation within the event or environment which is being researched (see Chapter 6 for an extended discussion). The most common form of qualitative geographical research involving participation is participant observation. Within participant observation, there may be a wide variation in the role of the observer from passive to pro-active (Hammersley and Atkinson 1983, p. 93). All forms of observation involve problems of positioning of the author in relation to the subject of the research (for example, Smith 1988). In particular, very active participation may clearly influence the event which is being researched, while researchers who are personally involved, for example by researching the community in which they grew up, may find it hard to wear their 'community' and 'researcher' hats at the same time. Participant observation allows the researcher to be, at least in part, simultaneously 'outsider' and 'insider', although differences in social status and background are hard to overcome (Moss 1995). The positioning of the researcher in relation to the 'researched' raises some significant ethical issues, especially if the research is covert (Evans 1988, pp. 207–08). It can, however, have some important advantages, for example for the student whose research work also provided him/her with paid work at a fast-food outlet (see Cook 1997 for an excellent account of this and other student projects). In-depth participant observation is essentially indistinguishable from **ethnographic approaches**, which often involve lengthy fieldwork. That fieldwork can enable meaningful relationships with the research subjects to develop and may facilitate deep understanding of the research context (Eyles 1988, p. 3; Cooper 1994; 1995). It is useful to follow Cook's (1997, p. 127) terminology where he states that participant observation is the means or method by which ethnographic research is undertaken.

The contribution of qualitative techniques to explanation in geography

In this chapter I have discussed two fundamental questions of geographic enquiry, those concerned with individuals and those concerned with social structures. I have also indicated three main groups of methods: the oral, the textual and the participatory. There is no simple relationship between the method used and the research questions posed. It is tempting to say that oral methods may be directed predominantly towards elucidating the experiences of individuals and their meanings; however, this is overly simplistic. People's own words do tell us a great deal about their experiences and attitudes, but they may also reveal key underlying social structures. In my own work on lone fathers, I found that the in-depth interviews illuminated underlying structures of patriarchy and masculinity in ways which were much more profound than anticipated (Winchester 1999). Depths of individual anger and despair reflected mismatches between those individuals' romanticised expectations of marriage and gendered behaviour and their actual experience of married life. In this sense, the oral method chosen elucidated both individual experiences and social structures in the holistic sense which would most frequently be associated with participant observation.

Similarly, it might appear that textual methods would most commonly be employed to throw light upon the social processes that underpin, legitimate and resist social structures. This generalisation would probably be more widely accepted than any equation of oral methods with research questions that focus on the individual. Textual methods have indeed been used to analyse some of the many social processes studied by contemporary human geographers. Examples that spring to mind include the discursive construction of place (Mee 1994; Dunn, McGuirk and Winchester 1995), processes of social exclusion (Duncan and Duncan 1988) and expressions of 'problematic' masculinity (Hartig and Dunn 1998).

One of the most recent areas of study in human geography focuses on the body and on our embodied experiences. A study by Longhurst (1995) of the experiences of pregnant women in shopping malls showed how the embodied experiences of individuals (of feeling marginalised, of needing more toilets, of being uncomfortable in particular places such as bars and lingerie shops) are indicative of the way the pregnant body is socially constructed as **'other'** (i.e., oppositional to or outside the mainstream) to be confined to particular places and roles, medicalised and marginalised. The study of the body may also be as a text or as a landscape which may be marked or shaped in particular ways either as a form of identity (McDowell and Court 1994 identify bodily performances of male and female merchant bankers) or as a form of resistance (Bell et al. 1994 commented on lipstick lesbians as practising resistance to heterosexual norms). The study of the body as text, as performance or as social construction, illuminates some of the richness of methods that cannot be easily pigeon-holed into the types of qualitative method and types of geographical explanation identified for convenience earlier in the chapter.

THE RELATIONSHIP BETWEEN QUALITATIVE AND QUANTITATIVE GEOGRAPHY

In the last two decades the pendulum of geographical methods within human geography has swung firmly from quantitative to qualitative methods. The two are generally characterised as in opposition or as conflicting methodologies. Qualitative methods have been in the ascendant since the 1980s. The trend towards the resurgence of qualitative sources and methods in geography has been chronicled in, and stimulated by, recent books and collections on qualitative methods and **mixed methods** (Brannen 1992b; Eyles and Smith 1988; Flowerdew and Martin 1997; Holland et al. 1991; Lindsay 1997).

Typically the gulf between qualitative and quantitative methods has been presented as a series of dualisms. Hammersley (1992) listed seven 'polar opposites' between qualitative and quantitative methods (Box 1.2). Similarly, Brannen (1992a) characterised qualitative approaches as viewing the world through a wide lens and quantitative approaches as viewing through a narrow lens. A dualistic view of methods is highly problematic, as Hammersley (1992, p. 51) recognised: it represents quantitative methods as focused, objective, generalisable and, by implication, value-free. On the other hand qualitative methods are often presented as soft and subjective, an anecdotal supplement, somehow inferior to 'real' science.

Box 1.2: Dualisms identified between qualitative and quantitative methods

Qualitative methods	*Quantitative methods*
Qualitative Data	Quantitative data
Natural Settings	Experimental settings
Search for meaning	Identification of behaviour
Rejection of natural science	Adoption of natural science
Inductive approaches	Deductive approaches
Identification of cultural patterns	Pursuance of scientific laws
Idealist perspective	Realist perspective

Source: after Mostyn (1985) and Hammersley (1992).

Such a view misleadingly represents quantitative methods as objective and value-free; increasingly this assumption about the nature of science has been questioned. Our choices of what we study and how we study it reflect our values and beliefs. For example much early feminist geography uncovered sexist assumptions in how geographers had typically studied and measured human behaviour (Monk and Hanson 1982). Measurements of migrants and shoppers which ignored 'half of the human population' were clearly demonstrated to be unobjective and value-laden, and in many cases strongly coloured by naturalised assumptions about gendered roles and behaviour. If the subjectivity and value-laden nature of all research methods is admitted, then the apparent gap between the two groups of methods is dramatically reduced. Geographers using qualitative methods often outline their personal subjectivity and possible sources of bias by summarising their own background as researchers and their relationship to the research and to its intended audience. Furthermore, the increasingly cultural 'turn' within human geography has made such openness more widespread across the discipline to include not only social and cultural geographers but those with more economic and political interests (for example, Thrift 1996). It is arguable that researchers who define their own position in relation to their research could be more objective than their colleagues who hide behind the supposed objectivity of quantitative methods without revealing the many subjective influences which shape both the research question and the explanations that they find to be true. In short, the equation of 'objectivity' with the quantitative and of subjectivity with the qualitative is highly contested (Philip 1998). This contest is discussed further in Chapters 2 and 9.

As qualitative methods have again become prominent within the discipline, they have increasingly had to be justified in a scholarly environment which had come to value measurement and scientific observation more highly than individual experience or social process. Within a largely unfriendly hegemonic scientific framework, advocates of qualitative studies have generally drawn from three arguments. First, some studies of individual experiences, places and events have been represented as essentially non-generalisable case-studies which have meaning in their own right but are not necessarily either representative or replicable (for example, Donovan 1988). The second argument, appropriate to some large-scale studies, has been to suggest that they have generated sufficient data to allow general, and sometimes quantified, conclusions to be drawn from their research (for example, Wearing 1984; Oakley 1981). More usually, however, qualitative methods have been justified as a complementary technique, as an adjunct or precursor to quantitative studies from which generalisations can be drawn, and as explorations in greater depth as part of multiple methods or triangulation (Burgess 1982a). In making these arguments, qualitative geographers have often been on the defensive, trying to present their studies as legitimate in their own right and as research which produces not just case-studies or anecdotal evidence, but which has added immensely to the geographical literature through powerful forms of geographical explanation, including analysis, theory building and geographic histories.

Classically, qualitative and quantitative methods, such as interviews combined with questionnaires, are seen as providing both the individual and the general perspective on an issue (for example, England 1993) while similar arguments have been raised for mixed methods more broadly (McKendrick 1996; Philip 1998). This triangulation of methods and use of multiple methods are sometimes deemed to offer cross-checking of results by approaching a problem from different angles and using different techniques. Brannen (1992a, p. 13), however, has argued that data generated by different methods cannot simply be aggregated, as they can only be understood in relation to the purposes for which they were created. This question of purpose is intimately related to the theoretical perspectives from which the techniques derive and is considered further in Chapter 3.

THE HISTORY OF QUALITATIVE RESEARCH IN GEOGRAPHY

Geography has existed as an academic discipline in Australia since the early years of the twentieth century (Gale 1996). Essentially, for the bulk of the time, geographical work has been dominated by qualitative research of a scholarly and informed, but unquantified nature, drawing assessments of evidence from both physical and human environments (for example, Powell 1988). It should be recognised that qualitative methods of many sorts have been used widely throughout this century, particularly in the development and writing of sensitive and nuanced regional geography, such as that of Oskar Spate on the Indian sub-continent (Spate and Learmonth 1967), in the landscape school of both human and physical geographers with 'an eye for country', but also in interviewing and field observation (Davis 1954; Wooldridge 1955). The postwar era of the 'quantitative revolution' may, in hindsight, be seen as an aberration rather than the revolutionary **paradigm** (mode of thought) that it was claimed to be at the time (Wrigley 1970).

Box 1.3: Paradigm shifts and research methods in geography during the twentieth century*

Time	*Paradigm*	*Research questions*	*Research methods*	*Characteristics*	*Trends*
Early 20th century ↑	**Exploration/Discovery**	Discovery	Exploration	Colonial	
	Classical Geography	General/Theoretical/ Contextual	Qualitative/Quantitative	Environmental determinism	Decrease in spatial and time scale
	Regional Geography	Unique/Empirical	Qualitative/Regional Delineation and description	Region both method and object of study	
↓	**Spatial Science**	General/Theoretical and Empirical	The 'Quantitative Revolution'	Scientific method	Increasing separation of physical from human geography
1980s +	**Critical Social Science**	Theoretical/Structural/ Individual	Qualitative	Pluralist	Increasing diversity in approaches
	Radical	Theoretical/Structural			
	Feminist	Structural/Individual			Rise of environmental studies
	Phenomenological	Unique/Individual	Qualitative	Local	
	Postmodern	Theoretical			
	Postcolonial/Subaltern	Theoretical/Empirical	Qualitative	Global and local	

***The shifts identified post 1980 are most relevant to human geography, rather than to geography in general.**

The early history of geographical thought has been represented classically as a series of paradigm shifts, each triggered by dissatisfaction with the previous prevailing paradigm. A schematic representation of paradigm shifts in academic geography is presented in Box 1.3 (Holt-Jensen 1988; Johnston 1983; Wrigley 1970). For example dissatisfaction with the crude environmental determinism of the early twentieth century prompted the study of unique places around the world. When this regional approach degenerated into stale layers of facts, and geographers had totally marginalised themselves from the academy by their commitment to 'the region'—both as object of study and research method—then an alternative, more credible to the academic community, was sought. The strategic alliances of the discipline shifted away from history and geology to newer and more innovative disciplines such as psychology and economics. By the 1960s, the quest for academic credibility, combined with a technological and data revolution, propelled more scientific ways of thinking into the discipline. This scientific approach combined the use of quantitative method, model making and hypothesis testing. The regional idiosyncrasies were condemned as 'old hat' and geographers turned themselves into spatial scientists. This very compressed 'history' has some validity for the earlier years of this century, although the notion of paradigm shifts has been challenged (for a concise review, see Gregory 1994).

The notion of paradigm shifts essentially becomes inapplicable in the confusing and exciting world of post-quantitative human geography (Billinge, Gregory and Martin 1984). It is recognised that in recent human geography there are coexistent, contradictory and competing communities of scholars adhering to different views of the world, different schools of thought and different approaches to research questions. Box 1.3 shows that the recent period is occupied by a number of competing viewpoints jostling for space and credibility. The reactions against normalising spatial science have spawned a huge diversity of approaches; by the early 1980s radical, feminist, and environmental geographers were reasserting the importance of the social, the agency of the individual, and the particularity of place. Qualitative research requiring qualitative methods reasserted its respectability.

From the schema outlined in Box 1.3, a few major points can be drawn:

1 The period of spatial science is unique and aberrant in focussing on quantitative methods.
2 The paradigm shifts within geography have involved an increasing separation in the methods and philosophies of human geography from those of physical geography. The reactions against 'scientific' geography established since the 1970s have drawn human geography more and more into the realms of critical social science, while physical geography has remained essentially within the scientific paradigm.
3 The questions that geographers have asked have oscillated between elucidating general trends and patterns in one period to examination of the individual and unique in subsequent phases of geographic enquiry. The plethora of post-quantitative approaches allows both individual and structural research questions to be tackled.
4 In general, the scale of geographic enquiry has shifted from the global, to the regional, to the local. However, recent writings are concerned not only with the specifics of individual experiences and places, but have re-engaged with both the theoretical and the global.

Qualitative methods are currently used by all the major groups within the critical social science approaches utilised in human geography and identified in the lower half of Box 1.3. Much of the drive for qualitative research has come initially from humanistic geography of the late 1970s, which focused geographers sharply on values, emotions and intentions in the search to understand the meaning of human experience and human environments (for example, Ley 1978). Another significant influence in the reassertion of qualitative methods has been the work of feminist geographers establishing links between the personal and the political. A clear example of this might be Valentine's (1989) work on the geography of women's fear or Mackenzie (1989) on women and environments in a postwar British city. This approach also predominates in studies of gay communities and environments (for example, Seebohm 1994). More recently, many studies which might be grouped as postcolonial give voice to people defined as 'other', enabling multiple interpretations of events to be heard. Much qualitative work within contemporary human geography cannot be clearly categorised within any of the schools of thought listed in the final section of Box 1.3 (critical social science), but is concerned with the broad questions of elucidating human environments and human experiences within a variety of conceptual frameworks.

CONTEMPORARY QUALITATIVE GEOGRAPHY—THEORY/METHOD LINKS

Contemporary human geography adopts a broad range of research methods. Although not tied specifically to particular theoretical and philosophical viewpoints, the methods discussed in the preceding sections of this chapter are often more frequently associated with one standpoint than another. For example feminist geography is often associated with qualitative methods through a naturalised association of the feminine with the 'softer' qualitative approaches. However, feminist questions can be stated within a variety of theoretical frameworks, and may use a variety of methods (Lawson 1995).

Qualitative methods have been used more widely in human geography throughout this century than is commonly believed. They have been used in conjunction with quantitative methods in a search for generality, and have also been used to explain difficult cases or to add depth to statistical generalisations; above all they have traditionally been used as part of triangulation or multiple methods in a search for **validity** and corroborative evidence. However, qualitative methods have also been used in different conceptual frameworks to reveal that which has previously been considered unknowable—feelings, emotions, attitudes, perceptions, and cognition. Overwhelmingly, qualitative methods have been used to verify, analyse, interpret and understand human behaviour of all types.

Studies that utilise qualitative methods in their own right to express individual meanings are much more limited in number. Although humanistic geographers of the 1980s laid claim to this territory, the output of the humanistic school *per se* was both limited and short-lived; even by 1981, Susan Smith was calling for 'rigour' in humanistic method in a way akin to the current calls for validity and **replicability** and (by implication) respectability (Baxter and Eyles 1997; Philip 1998).

Similarly the aims of critical realism expressed by Sayer and Morgan (1985) have given pre-eminence to individuals' actions and their meanings, yet contributions in this mould to geography literature have been slim. The schools of thought that may have made the greatest contribution to answering qualitative research questions have been the feminist and the poststructural (including the postmodern and the postcolonial). These frameworks both recognise that multiple and conflicting realities coexist. They deliberately give voice to those silenced or ignored by hegemonic (modern, colonial) views of histories and geographies. They embody and acknowledge previously anonymous individuals. Paradoxically, however, the voice of the oppressed not only speaks for itself: it is part of a wider whole. Reality is like an orchestra: poststructural approaches differentiate the instruments and their sounds and bring the oboe occasionally to centre-stage; usually dominated by the strings, the minor instruments too have a tune to play and a thread that forms a distinct but usually unheard part of the whole. It is the voices of the women and children, the colonised, the indigenous, the minorities, which, when released from their silencing, enable a more holistic understanding of society to be articulated (for example, Jane Jacobs' [1993] interpretation of the Aboriginal and mining representations of the conflict over Coronation Hill).

Sayer and Morgan (1985) make the point that exactly the same research technique can be used in different ways for different purposes, according to the theoretical stance of the researcher. Interviews, for example, may be used to access information from gatekeepers about structures or to give voice to silenced minorities. John McKendrick (1996, Table 1) considers the relationship between methods and their applications in different research traditions. He contrasts the use of interviews in the humanistic tradition, to 'explore the meaning of the migration of each individual migrant', with their use in a postmodernist framework, whereby in-depth interviews with women may be used 'to "unpack" their rationalisations of their migrations'.

Qualitative methods raise an immense number of difficult issues that are considered in more detail in several of the chapters of this volume. Among those issues are concerns over authorship, audience, language, and power. Ethnographic research is often highly complex, within which the individual subject and the audience for the research are often intermingled and mutually dependent. The position of the author as observer in relation to the object of research raises issues of power relations and control. The engagement of the researcher does not necessarily allow the voices of the researched to speak as they are mediated through the researcher's experience and values. The language of research reporting may also exclude those researched, although language varies according to the audience towards which the research is directed. The key issue of the outsider gazing, perhaps voyeuristically, at those defined as 'other' is an intractable problem that needs to be recognised even if it cannot be solved. Even participant observation cannot surmount inbred and naturalised class differences, as demonstrated by Moss (1995) in her immersion in manual labour in hotels. Moss never managed to bridge the cultural, linguistic, and social gap between herself as middle-class researcher and the housemaids who spent their lives in manual labour. The engagement with human research subjects raises significant questions about ethical research practice which are only now being addressed seriously within the discipline.

SUMMARY

This chapter has outlined three major groups of qualitative methods currently used in human geography. These are oral, textual and participatory methods. The range of methods is used to answer two broad research questions, relating either to the experiences of individuals or to the social structures within which they operate. Qualitative methods have often been categorised as oppositional to quantitative methods, yet in many respects this is a false dichotomy. The differences between qualitative and quantitative methods relate to the conceptual frameworks from which they have been derived. In elucidating human experiences, environments and processes, qualitative methods attempt to gather, verify, interpret and understand the general principles and structures that quantitative methods measure and record. Furthermore, qualitative methods have very frequently been used in conjunction with other methods. The use of qualitative methods alone to explore human values, meanings, and experiences has been more limited. Currently some of the most rewarding qualitative research in human geography is being carried out in feminist and postcolonial frameworks to enable silenced voices to be heard and to foster better comprehension of those naturalised discourses which exclude and marginalise certain groups.

KEY TERMS

deduction
discourse
epistemology
ethnographic approaches
humanism
individual experiences
induction
interviews/interviewing
landscape
mixed methods
objective
ontology
'other'
oral methods
paradigm
participation/participant observation
postcolonial
qualitative methods
quantitative methods
replicability
semiotics
structures/social structures
subjective
text/textual
triangulation
validity

REVIEW QUESTIONS

1. What research questions may be answered by qualitative methods? Give examples.
2. What are the main types of qualitative research methods used in human geography?
3. How may different types of qualitative methods be linked to theoretical approaches within the discipline?
4. Outline what is meant by the quantitative/qualitative debate.

SUGGESTED READING

Anderson, K. J. 1995, 'Culture and nature at the Adelaide Zoo: at the frontiers of "human" geography', *Transactions of the Institute of British Geographers*, vol. 20, no. 3, pp. 275–94. This article examines discourses surrounding the 'natural' world.

Cook, I. 1997, 'Participant observation', in *Methods in Human Geography*, eds R. Flowerdew and D. Martin, Addison Wesley Longman, Harlow. This chapter contains many useful hints and stories which make it very useful for students thinking about starting projects. It focuses on the third type of method discussed in this chapter (i.e., participatory).

Jackson, P. and Holbrook, B. 1995, 'Multiple meanings: shopping and the cultural politics of identity', *Environment and Planning* A, vol. 27, pp. 1913–30. This article employs focus groups and provides an example of the 'deconstruction' of an everyday activity.

Kong, L. 1999, 'Cemeteries and columbaria, memorials and mausoleums: narrative and interpretation in the study of deathscapes in geography', *Australian Geographical Studies*, vol. 37, no. 1, pp. 1–10. This article analyses landscape, described in this chapter as the second main type of qualitative method.

Longhurst, R. 1995, 'The geography closest in—the body.... the politics of pregnability', *Australian Geographical Studies*, vol. 33, pp. 214–23. A highly readable article which focuses on embodied experience and the discourses surrounding embodiment, but which is grounded in the reality of women's shopping experiences.

Winchester, H. P. M., McGuirk, P. M., and Everett, K. 1999, 'Celebration and control: Schoolies Week on the Gold Coast Queensland', in *Embodied Geographies: Spaces, Bodies and Rites of Passage*, ed. E. Teather, Routledge, London. This chapter uses oral qualitative evidence to examine a cultural phenomenon of a rite of passage. It provides an example of oral methods.

2

Power, Subjectivity and Ethics in Qualitative Research

Robyn Dowling

CONTENTS

CHAPTER OVERVIEW

This chapter aims to introduce issues that arise because qualitative research typically involves interpersonal relationships, interpretations and experiences. I discuss three issues of which qualitative researchers need to be aware: (1) the more formal ethical issues raised by qualitative research projects; (2) the power relations of qualitative research; and (3) objectivity, subjectivity and intersubjectivity. Rather than advocating simple prescriptions for dealing with these issues, the chapter proposes that researchers be 'critically reflexive'.

INTRODUCTION: ON THE SOCIAL RELATIONS OF RESEARCH

Chapter 1 outlined the types of research questions asked by qualitative researchers, namely our concerns with the shape of societal structures and people's experiences of places and events. This chapter takes you one step closer to conducting qualitative research. It discusses some of the implications of research as a social process. Collecting and interpreting social information involves personal interactions. Interviewing, for example, is essentially a conversation, albeit one contrived for research purposes. Interactions between two or more individuals always occur in a societal context. Societal norms, expectations of individuals, and structures of power influence the nature of those interactions. For instance, when you are conducting a focus group, you may find men talking more than women, and people telling you what they think you want to hear. Societal structures and behaviours are not separate from research interactions. This places all social researchers in an interesting position. We may use a variety of different methods to understand society, but those methods cannot be separated from the structures of society. The converse is also true. The conduct of social research necessarily has an influence on society and the people in it. By asking questions or participating in an activity we alter people's day-to-day lives. And communicating the results of research can potentially change social situations.

Both qualitative and quantitative researchers recognise this lack of separation between research, researcher and society. What distinguishes qualitative researchers' approach to this issue is the emphasis they give to it. For those who use the qualitative methods discussed in this volume, the interrelations between society, the researcher, and the research project are of critical and abiding significance. They permeate all methods and phases of research. These relationships cannot be ignored, and raise key issues that must be considered when designing and conducting research. This chapter outlines three issues that arise because of the social nature of research and suggests an approach to dealing with them.

The chapter begins with a discussion of ethical guidelines commonly applied to research projects. It moves beyond these guidelines to introduce the concept of critical reflexivity. The chapter then focuses on the ways power traverses the conduct of qualitative methods. Finally, the chapter considers the significance a researcher's subjectivity and intersubjective relations have for the collection of qualitative data.

A word of caution to begin. In general, subsequent chapters offer practical guides to different qualitative methods. They explain how to be a participant observer, how to read a text, how to conduct a focus group, and so forth. This chapter is different in two important respects. First, it is not about any specific method. It is about issues common to each of the methods discussed in this volume. This chapter should be used in combination with those on particular methods to help you think about the specific challenges you are likely to encounter in your research. Second, this chapter does not—and cannot—offer hard and fast rules on conducting ethical research that is responsive to matters of power and **intersubjectivity**. The conduct of good, sensitive, and ethical research depends, in large part, on the ways you deal with your unique relationships with research participants at particular times in particular places.

UNIVERSITY ETHICAL GUIDELINES

Ethics, broadly defined as being about 'the conduct of researchers and their responsibilities and obligations to those involved in the research, including sponsors, the general public and most importantly, the subjects of the research' (O'Connell, Davidson and Layder 1994, p. 55), constitute an issue that must be dealt with in your research. University ethics committees focus on the researcher's responsibilities to research subjects, and formulate guidelines about what researchers should and should not do. It is useful to consider such formal guidelines as a first step in thinking through the social context of your research. In most Australian universities honours and postgraduate students are required to gain the formal approval of a university ethics committee before beginning their research, while lecturers are typically responsible for gaining approval for undergraduate projects. The committee will not evaluate your research design, but will want to know the aims of the research and the methods you will use. It will be concerned primarily with your responsibilities to research participants with regard to matters of privacy, informed consent and harm.

Privacy and confidentiality

Qualitative methods often involve an invasion of someone's privacy. You may be asking very personal questions or observing interactions in people's homes that are customarily considered private. Ethics committees are concerned that these private details about individuals are not released into the public domain. Accordingly, you may have to show that your original fieldnotes, tapes and transcripts will be stored in a safe place where access to them will be restricted. You may also need to ensure that your research does not enable others to identify your informants. There are various ways of ensuring the anonymity of informants, including using pseudonyms and masking other identifying characteristics (for example, occupation, location) in the written version of your research. You should note, however, that when dealing with significant public figures it is sometimes not possible or desirable to ensure anonymity. For example, if your interest is in the responses of BHP (Broken Hill Proprietary Company Limited) to economic restructuring and you interview the corporation's Chief Executive Officer about that matter, you may wish to ask for permission to use their name in public reports of your work.

Informed consent

For most geographical research, participants must consent to being part of your research. In other words, they have to give you permission to involve them. However, this criterion is somewhat stricter than a simple 'yes, you can interview me'. It must be ***informed* consent**. Informants need to know exactly what it is that they are consenting to. You need to provide participants with a broad outline of what the research is about, the sorts of issues you will be exploring, and what you expect of them (for example, the amount of time required to complete an interview). Most ethics committees recognise that there are exceptions to informed consent. Simple observation of people in a place like a public shopping mall, for example, may not need the explicit consent of those individuals. Indeed, it may be physically impossible to secure the consent of everyone involved. Sometimes informed consent may be waived, although an ethics committee will typically ask you to justify that decision. There are some relatively rare instances when research may involve deception. Deception is where research participants either do not know that you are a researcher (for example, Chapter 6 on participant observation) or do not know the true nature of your research. Set against the principle of informed consent, deception is clearly an ethically difficult issue and you should think carefully and seek advice before contemplating a research project that involves deception. Moss' (1995) discussion of why and how she used deception in her study of domestic workers may be useful.

Harm

Your research should not expose yourself or your informants to harm—physical or social. As social scientists it is highly unlikely that you will be subjecting people to physical harm. You may, however, be bringing them into contact with 'psycho-social' harm. You may raise issues that may be upsetting or potentially psychologically damaging. This does not mean that your research cannot proceed. Rather, it means your research should cater for this possibility. For example, in work with gay men in western Sydney, Stephen Hodge (reported in Costello and Hodge 1998) had ready the contact details of a counselling service in case participants became upset during interviews. You should also avoid putting yourself at risk during the research. For example, a young woman planning a participant observation project on single women's safety at night on public transport could meet a very cautious response from an ethics committee or research supervisor because of the potential dangers to herself while conducting that work.

MOVING BEYOND ETHICAL GUIDELINES: CRITICAL REFLEXIVITY

Although important, ethical rules and ethics committees are not unproblematic. Iain Hay (1998), for instance, argues that rigid codes cannot deal sometimes with 'the variability and unpredictability of geographic research' (p. 65). What is appropriate in one situation will be inappropriate in another. The blanket application of rules for informed consent, for example, does not take account of the specificity of individual circumstances and character of some research projects (for example, observing people in a public park or plaza). More relevant to this chapter is the suggestion that as geographers engaged in research, we must constantly consider the ethical implications of our activities. Because research is a dynamic and ongoing social process that constantly throws up new relations and issues that require constant attention, self-critical awareness of ethical research conduct must pervade our research. Our engagement with ethical behaviour does not end when we submit our research proposal to an ethics committee.

As the foregoing discussion of ethical research conduct might imply to you, human geographers have come to appreciate more fully than ever before the social nature and constitution of our research. Indeed, we now recognise and acknowledge this location through the concept of ***critical reflexivity***. Reflexivity, as defined by Kim England (1994), is a process of constant, self-conscious, scrutiny of the self as researcher and of the research process. In other words, being reflexive means analysing your own situation as if it were something you were studying. What is happening? What social relations are being enacted? Are they influencing the data?

Critical reflexivity is difficult but rewarding. It is rewarding in that, as some of the examples used in this chapter indicate, it can initiate new research directions. Critical reflexivity is however difficult in two respects. First, many geographers do not write about the research process in their published work. Linda McDowell (1998) comments, for example, that many of the details of how her merchant banking research proceeded and, as a result, her reflections on the process, do not appear in the book based on that research. Second, reflexivity is difficult because we are not accustomed to examining our engagement with our work with the same intensity as we regard our research subjects. You may be helped in this matter by keeping a **research diary**, which is outlined in Box 2.1 below.

Box 2.1: The research diary as a tool for the reflexive researcher

Your task in being reflexive will be helped by keeping a research diary. The contents of a research diary are slightly different from those of a **fieldwork diary**. While a fieldwork diary, or fieldnotes, contains your qualitative data—including observations, conversation, and maps—a research diary is a place for recording your reflexive observations. It contains your thoughts and ideas about the research process, its social context and your role in it. You could start your research diary by including answers to the questions posed in the checklist set out in Box 2.3 at the end of the chapter.

The rest of the chapter focuses on two of the important issues about which you need to be reflexive: power and subjectivity. For those readers who are especially interested, the discussion of reflexivity is extended in Chapter 9.

POWER RELATIONS AND QUALITATIVE RESEARCH

One important outcome of the social character of qualitative research is that research is also interleaved with relations of power. Power intersects research in a number of ways. It can enter your research through the stories, or interpretations, you create from the information you gather. Power is involved here because knowledge is both directly and indirectly powerful. Knowledge is directly powerful through its input into policy. Some studies are specific analyses of policy issues and their results have a direct impact on people's lives. Knowledge is also indirectly powerful. The stories you tell about your participants' actions, words and understandings of the world have the potential to change the way those people are thought about. Power is also involved in earlier parts of the research process. In undertaking qualitative research you are attempting to understand, participating in, and sometimes creating situations whereby people (yourself included) are differently situated in relation to social structures. Both you and your informants occupy different 'speaking positions'. Not only do you and your informants have different intentions and social roles, but you also have different capacities to change situations and other people.

England (1994) identifies the different sorts of power relations typically entered into by social researchers. She divides these relationships into three types. ***Reciprocal* relationships** are those whereby the researcher and the researched are in comparable social positions and have relatively equal benefits and costs from participating in the research. You may, for example, be conducting focus groups with your fellow students on how they are adjusting to university life. Although not absent, power differences in this relationship are minimal compared with two other sorts of relationships. In ***asymmetrical* relationships** those being studied are in positions of influence in comparison to the researcher. Interviews with corporate executives may fall into this category because of such executives' relative access to cultural and financial resources (see McDowell 1992). In ***potentially exploitative* relationships** the researcher is in a position of greater power than the research participant. Hilary Winchester and Lauren Costello's (1994) observations on the social networks of homeless youth provide an example of this.

It is in recognising and negotiating relations of power (and ethics) that critical reflexivity becomes important. It is easier to develop a response for some issues than it is for others. For example, when collecting data, our responsibility to the research participants is such that we should not take advantage of someone's less powerful position to gather information. In the case of homeless youth, for example, you would not make the possibility of gaining access to shelter dependent on that person's participation in the study. But you cannot eliminate the power dimension from your research, since it exists in all social situations. The best strategy is to be aware of, understand, and respond to it in a critically reflexive manner. Critical reflexivity does not necessarily mean altering your research design but it does imply that you reflect constantly on the research process and modify it where appropriate. When you are formulating your topic think about the various ethical and power relations that may be enacted during your research (see, for example, Box 2.2). Are you happy with the situation? Would you like to do anything differently? Could you justify your actions to others? You should also think about how you communicate the results. Have you reflected as faithfully as possible what you have been told and/or observed without reproducing stereotypical representations? Are you presenting what you heard and saw, or what you expected to hear and see? Remember, the stories you tell may change the worlds in which you and your research participants live (for more detail on this matter, see Chapter 9).

Box 2.2: Sexism in research

Sexism can present problems in many different sorts of research projects. Eichler (1988) identifies four primary problems of sexism in research:

- *androcentricity/gynocentricity*—a view of the world from male/female perspectives respectively. For instance, concepts of 'group warfare' developed through reference to men's experiences only.
- *overgeneralisation*—a study is only about one sex but presents itself as applicable to both sexes. By way of example a study that is exclusively concerned with men's location decisions might be misleadingly entitled 'Residential location decisions in Auckland'.
- *gender insensitivity*—ignores gender as an influential factor in either the research process or interpretation. For instance, a study of the geographical effects of a free-trade agreement that fails to consider any gender-specific effects.
- *double standards*—identical behaviours or situations are evaluated, treated or measured

by different means or criteria (for example, drawing different conclusions about men and women on the basis of identical answers to a survey or aptitude test).

Eichler also identifies three 'derived' forms of sexism:

- *sex appropriateness*—the notion that some characteristics and behaviours are accepted as being more appropriate for one sex than the other (for example, designing a research project on parental perceptions of children's play space, and interviewing women only because you assume they will know more about the issue).
- *familism*—using family as the unit of analysis when the individual might be more appropriate or vice versa (for example, working at the family scale in evaluating the social costs and rewards of in-home care for the elderly, rather than exploring the different implications this might have for males and females within the family).
- *sexual dichotomism*—postulating absolute differences between women and men (for example, women are sometimes claimed to be 'naturally' more timid than men).

OBJECTIVITY, SUBJECTIVITY AND INTERSUBJECTIVITY IN QUALITATIVE DATA COLLECTION

Objectivity has traditionally been emphasised in geographic discussions of quantitative research methods. Objectivity has two components. The first relates to the personal involvement between the researcher and other participants in the study. The introduction to this chapter suggested that it is impossible to achieve this sort of objectivity because of the social nature of all research. Objectivity's second component refers to the researcher's independence from the object of research. This implies that there can be no interactive relationship between the researcher and the process of data collection and interpretation (see Chapters 6 and 9). Clearly, however, dispassionate interpretation is difficult if not impossible because we all bring personal histories and perspectives to research. **Subjectivity** involves the insertion of personal opinions and characteristics into research practice. Qualitative research gives emphasis to subjectivity because the methods involve social interactions. As will become evident in later chapters, you need to draw on your personal resources to establish rapport and communicate with informants. Reading texts and landscapes also involves your subjectivity in that your everyday understanding of the world helps you decipher the texts. If subjectivity is important then so too is intersubjectivity. This refers to the meanings and interpretations of the world created, confirmed or disconfirmed as a result of interactions (language and action) with other people within specific contexts. Collecting and interpreting qualitative information relies upon a dialogue between you and your informants. In these dialogues your personal characteristics and social position—elements of your subjectivity—cannot be fully controlled or changed because such dialogues do not occur in a social vacuum. The ways you are perceived by your informants, the ways you perceive them, and the ways you interact are at least partially determined by societal norms.

Critical reflexivity is the most appropriate strategy for dealing with issues of subjectivity and intersubjectivity. While you cannot be entirely independent from the object of research, trying to become aware of the nature of your involvement, and the influence of social relations, is a useful beginning that can help you identify the implications of subjectivity and intersubjectivity in your research.

Geographers' work on gender provides some good examples of the role of intersubjectivity in research. Gender is important because we often ascribe characteristics to people on the basis of gender. Furthermore personal interactions vary with the gender of participants; we tend to react differently to men and women. Therefore gender is a factor which can influence data collection. For instance, Andy Herod (1993) found that male union officials restricted the sort of information he was given during interviews because he was a man. Specifically, his informants downplayed the role of women in the union's struggles. Herod attributes this not only to the perspectives of the union officials, but also to the social (masculine) context of the interviews. The union officials assumed, Herod thinks, that either he was not interested in gender or that it was inappropriate to raise this issue in the context of a male-to-male conversation. By contrast, Hilary Winchester (1996) suggests that being a woman interviewing men aided her research on lone fathers. She found herself adopting a typically feminine role of facilitating conversation with men, which helped considerably in gathering the men's stories.

Your ability to interpret certain situations also depends on your own characteristics. Important here is a debate about the relative merits of being an '**insider**' or '**outsider**'. An insider is someone who is similar to their informants in many respects, while an outsider differs substantially from their informants. As a white, non-immigrant academic interviewing migrant women, Thompson (1994) was an outsider in her research, while Mee's (1994) personal experiences in western Sydney make her research on images of western Sydney 'insider research'. One position in the debate is that as an insider both the information you collect and your interpretations of it are more valid than those of an 'outsider'. People are more likely to talk to you freely, and you are more likely to understand what they are saying because you share their outlook on the world. If you are not a member of the same social group as your informants, then establishing rapport may be more difficult (see Chapter 4). And, since you do not share their perspective on the world and their experiences, then your interpretations may be less reliable. The strongest claim for insider research comes from those contemplating cross-cultural research (for example, a *pakeha* ['white'] New Zealander conducting work with Vietnamese immigrants). But this is far from an accepted position in human geography. It has been argued (England 1994) that being an outsider can bring benefits to the research. It may mean that people make more of an effort to clearly articulate events, circumstances and feelings to the researcher. Kim England (1994), for example, believes that being a young woman helped her get information from senior executives in some of her research work.

A final perspective to consider is that you are never simply either an insider or an outsider. We have overlapping racial, socio-economic, gender, ethnic and other characteristics. If we have multiple social qualities and roles, as do our informants, then there are many points of similarity and dissimilarity between ourselves and research participants. Indeed, becoming aware of some of these commonalities and discrepancies can be one of the pleasures and surprises of qualitative research. Richards (1990), for example, found that her age unexpectedly worked in her favour in a study of suburban living. She had initially thought that where she lived, her family situation and her educational background would make it difficult to speak with and interpret the lives of suburban residents. On the contrary, her age (under 40) and life stage (with young children)—which she had in common with many residents—made speaking with the residents easier than she had anticipated.

Intersubjectivity also means that neither yourself, your participants, nor the nature of your interactions will remain unchanged during the research project. Your general outlook and opinions may indeed change as result of your research. If you are a participant observer, for example, you will be immersing yourself in a situation that will invariably affect you. Robin Kearns (1997) discusses how both he and his research project changed in response to his interactions as an observer. Kearns began his project on Maori health in rural New Zealand in the waiting room of a medical clinic. He did not intend to interact with patients in the waiting room. His attempts to be unobtrusive failed due to the inquisitiveness of patients who talked to him incessantly. These interactions had two immediate effects. First, they forced him to adopt a different stance in his research: he had to be more open about his presence and his research. Second, he modified his method by actively involving the local community in the research design and the findings (Kearns discusses this more fully in Chapter 6).

Critical reflexivity can give rise to new and exciting directions in one's research. In a very different context, Gilbert (1994) began a research project believing that the women she was speaking with were not feminists. But responses such as 'she had devoted her life to God because she was not going to let any man stop her from surviving' (p. 92) helped challenge her own preconceived ideas about feminism, and shed new light on the empirical material.

SUMMARY AND PROMPTS FOR CRITICAL REFLEXIVITY

Using the qualitative methods described in subsequent chapters will involve you in various social relations and responsibilities. I have advocated critical reflexivity—self-conscious scrutiny of yourself and the social nature of the research. Critical reflexivity means acknowledging rather than denying your own social position and asking how your research interactions and the information you collect are socially conditioned. How, in other words, are your social role and the nature of your research interactions inhibiting or enhancing the information you are gathering? This is not an easy task since it is not always possible to anticipate or assess accurately the ways in which our personal characteristics affect the information we accumulate.

I shall not end this chapter with answers. Instead, I offer a preliminary set of prompts that might help you reflect critically on issues of ethics, power and intersubjectivity in different types and phases of research (see Box 2.3).

Box 2.3: How to be critically reflexive in research

Before beginning:

- What are some of the power dynamics of the general social situation I am exploring and what sort of power dynamics do I expect between myself and my informants?
- In what ways am I an insider and/or outsider in respect to this research topic? What problems might my position cause? Will any of them be insurmountable?
- What ethical issues might impinge upon my research (for example, privacy, informed consent, harm, coercion, deception)?

After data collection:

- Did my perspective and opinions change during the research?
- How, if at all, were my interactions with participants informed/constrained by gender or any other social relations?

Remember to take notes throughout data collection and keep them in a research diary.

During writing and interpretation:

- Am I reproducing racist and/or sexist stereotypes? Why and how?
- What social and conceptual assumptions underlie my interpretations?

KEY TERMS

asymmetrical power relation
critical reflexivity
ethics
fieldwork diary
informed consent
'insider'
intersubjectivity
objectivity
'outsider'
potentially exploitative power relation
reciprocal power relation
research diary
subjectivity

REVIEW QUESTIONS

1. What is critical reflexivity? Read a study in human geography in your field of interest which uses qualitative methods. Is the researcher reflexive about the research process? If so, how? If not, what sort of questions would you like to ask the researcher about the research process?
2. What forms of power relations may be part of a qualitative research project?

3 Read a piece of qualitative research in human geography. Can you identify any or all of Eichler's (1988) forms of sexism in the way the research has been carried out or in the interpretation?
4 How are qualitative methods intersubjective? How might social relations like gender influence the collection of data?
5 Outline and explain the issues of concern to university ethics committees. Find out how your university's committee deals with these issues.

SUGGESTED READING

University web sites often contain ethical guidelines. Macquarie University's web site contains both special guidelines concerning Aboriginal and Torres Strait Islander peoples and a useful bibliography (http://www.ro.mq.edu.au/HEthics/).

Iain Hay discusses the limitations of ethical codes and suggests alternatives in a 1998 article entitled 'Making moral imaginations: research ethics, pedagogy, and professional human geography', *Ethics, Place and Environment*, vol. 1, no. 1, pp. 55–75.

Gillian Rose provides a sophisticated discussion of critical reflexivity and some of its problems in her 1997 article 'Situating knowledges: positionality, reflexivities and other tactics', *Progress in Human Geography*, vol. 21, pp. 305–320.

3

Qualitative Research Design and Rigour

Matt Bradshaw and Elaine Stratford

CONTENTS

CHAPTER OVERVIEW

Careful design and rigour are crucial to the dependability of any research. Research that is poorly conceived results in research that is poorly executed, and in findings that do not stand up to scrutiny. Thoughtful planning of research and the use of procedures to ensure that research is rigorous should therefore be central concerns for qualitative researchers. The research questions we ask, the cases and participants we involve in our studies, and the ways we ensure the rigour of our work all need to be considered in any dependable research.

INTRODUCTION

In this chapter, we focus on some matters of design and **rigour** that qualitative researchers need to consider throughout a project to ensure that the work satisfies its aims and its critical 'audiences'. We outline various principles of qualitative research design as well as some specific means by which rigour can be achieved in our work.

The chapter is organised into three main sections. First, we discuss influences on us as researchers, as well as the influence we have over the conduct of research. This discussion makes a link between the interpretive communities in which we work and the sorts of issues that are raised when we begin a research project. Second, we elaborate on how to select suitable cases and participants for study. In qualitative research, the number of people we interview, communities we observe, or texts we read is less important than the quality of who or what we involve in our research, and how we conduct that research. Third, we outline some ways of ensuring rigour in qualitative research to produce work that is dependable.

Careful research design is an important part of ensuring rigour in qualitative research and while texts and topics on research methods and design often imply that studies should be conducted in a particular 'right' way (Gould 1988), no single correct approach to research design can be prescribed. For certain kinds of work, the order and arrangement of stages may be different, stages can overlap, other stages might well be included, and the combination of qualitative and quantitative research is also possible. Nevertheless, by the end of the chapter we will have moved through various stages of qualitative research design, and summarised this passage in three diagrams. We believe that you will find this movement helpful in approaching your own qualitative research work.

ASKING RESEARCH QUESTIONS

Each of us needs to acknowledge that our fellow geographers and other colleagues are already involved in our studies (Box 3.1). We do not formulate research questions and undertake research in a vacuum. We are all members of **interpretive communities** that involve established disciplines with relatively defined and stable areas of interest, theory, and research methods and techniques (Fish 1980; Butler 1997). Our interpretive community influences our choice of topic, and our approach to and conduct of study. We also fold our own values and beliefs into research, and these can influence both what we study and how we interpret our research (see Jacobs 1999 and Chapter 9 of this volume for more detail).

Box 3.1: Considerations in research design, stage 1: asking research questions*

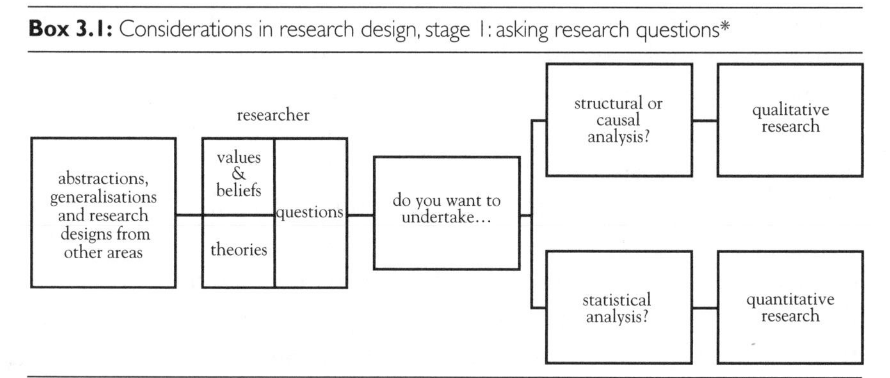

*The division between qualitative and quantitative research in this diagram and in Box 3.2 is simply an aid to simplify presentation. It does not signify that the two approaches are mutually exclusive.

From asking research questions to conducting research

Research aims affect research design. For example, asking 'how many skaters frequent a particular public place compared with other types of users?' will involve a research design different from one that aims to answer the question '*why* do skaters use this place and how do they interact with other types of users?' The first question focuses on quantification and statistical analysis, while the second is more concerned with the qualitative investigation of underlying causes of skaters' behaviour and the ways this partly expresses various structural relations.

In considering the ways in which we will conduct our research, we need also to ask what we want to do with the information we collect. Answering this question will also help us to decide what kind of research to do. Before making this decision, we need to be aware of some of the differences between qualitative (intensive) and quantitative (extensive) research. As Sayer (1992) suggests, each method helps us to answer different research questions, employs different research methods, has different limitations, and ensures rigour differently. Extensive research is characterised by the identification of regularities, patterns and distinguishing features of a population, often through a sample that has been selected using a random procedure to maximise the possibility of generalising to a larger population from which it is drawn. Extensive methods are designed to establish statistical relations of similarity and difference among members of a population, but they lack explanatory power. For instance, we may be able to determine that N number of respondents in a sample think P, but we may not readily be able to determine *why* it is they hold this opinion. Intensive methods require that we ask how processes work in a particular case (Platt 1988). We need to establish what actors do in a case, why they behave as they do, and what produces change both in actors and in the contexts in which they are located.

For example the issue of multiple-use conflict in a public place can be investigated using quantitative methods. Wood and Williamson (1996) decided to distribute a standardised questionnaire to a random sample of the users of Franklin Square, an inner-city square in Hobart, Tasmania that had been partly claimed through day-to-day use by skaters. The data from their study were aggregated and statements were made about the degree to which these data were likely to reflect the opinions of all the Square's users about the presence of skaters. This extensive approach produced useful information suggesting the existence of common characteristics and patterns; for instance, skaters used certain parts of the Square, while other users avoided these. But such findings did not account for the shifting quality of various people's *different* experiences of Franklin Square and of each other, or the reasons behind their opinions. Also selecting Franklin Square as a case by which to examine multiple-use conflict in public places, Stratford (1998) has used intensive methods such as in-depth interviews and observation to understand various responses to skating in the Square. If we are interested in working through the elements of structure and process that arise from analysing responses rather than in data that make statistical analysis possible, then we are pointed in the direction of intensive research.

In summary, then, in moving towards a qualitative research design, we are influenced by the theories we are concerned to use, by studies undertaken by other researchers in our interpretive communities that we have found interesting, and by the research questions we wish to ask—all of which are interrelated.

SELECTING CASES AND PARTICIPANTS

The ways in which the terms '**case**' and '**participant**' are defined will differ between interpretive communities. However, it is our view that cases are examples of more general processes or structures that can be theorised. Researchers should be able to ask 'that categorical question of any study: "What is this case a case of?"' (Flyvbjerg 1998, p. 8). Franklin Square is a *case* of multiple-use conflict in a public place, with this more general process involving, for example, theories of consumption, citizenship and government. *Participants* make up some of the elements of the case in question, for example, skaters, the elderly, the business community and Council.

Selecting cases

Sometimes we find a case, and sometimes a case finds us. In both instances, selection combines purpose and serendipity (Box 3.2). On the one hand, we may read about multiple-use conflict in public places in other cities, and want to see if causal explanations advanced there have merit in—or inform our understanding of—situations with which we are familiar. In this instance, the general or theoretical interest 'drives' the research and we must narrow the field, selecting cases and participants for research. On the other hand, perhaps a local government Parks Manager draws our attention to conflict among various groups in one site in the city, and wishes us to investigate options to manage this conflict. In this situation, the case has 'found' the researcher—and theories about multiple-use conflict in public spaces are subsequently woven into it. It is worth noting that if, for example, a Community Development Officer rather than a Parks Manager had contacted us about the same issue, we may well have been presented with a different brief which would in some ways make for a 'different' case.

Irrespective of how a case is selected, it is usually advisable to work in sites or on cases that are both practical and appropriate. In our example of skaters' use of public places, ambiguous sites—such as shopping malls, which are generally private places behind public façades—may need to be eliminated. In practical terms, we must be able and be permitted to work in the site or sites we select.

Box 3.2: Considerations in research design, stage 2: selecting cases and participants

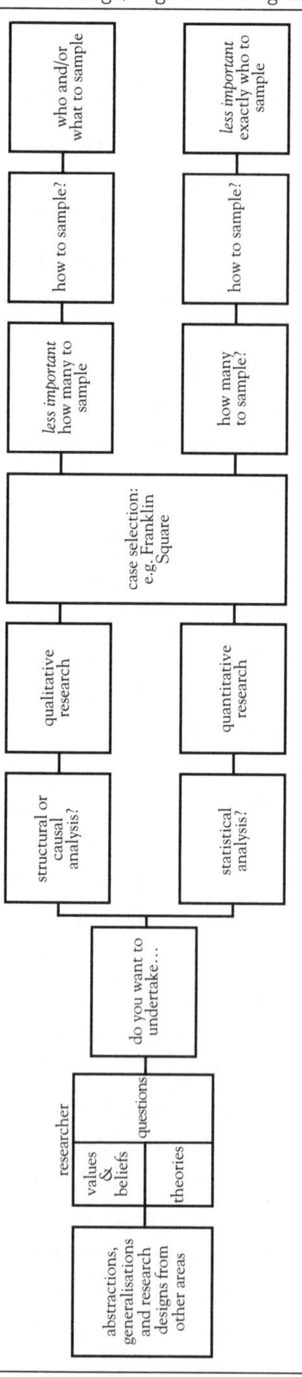

One final issue needs to be considered in case selection. On the one hand, we might choose to work with cases that are deemed 'typical' on the grounds that these will provide useful insights into causal processes in other contexts. Alternatively, we might deliberately seek out **disconfirming** cases. Such cases might include individuals or observations that challenge a researcher's interpretations or do not confirm ways in which others portray an issue. It might be, for example, that we have studied media reports in which it is suggested that there is unmitigated conflict between youths and the elderly in a public square. However, interviews with elderly pensioners lead us to understand that certain aged people regularly frequent the square at the same times as youths because they seek to be among the young, whose company they find enjoyable. Such disconfirming cases can be important in the research process, requiring us to think through how different institutions and the practices used by them (such as the media and their tendency to sensationalise events) create stereotypes. Such cases also require us to ask how various actors are represented (and for what reasons) and how they represent themselves.

Selecting participants

Generally speaking, the more focused our research interest becomes, and the better our background information and understanding, the more certain we are about who we wish to involve in our research and why. Exploratory work (for example, reading, observation, viewing television documentaries, conducting preliminary interviews) will often give us the capacity to begin to comprehend the perspectives of key informants. Understanding key informants in complex cultural situations usually requires semi-structured, in-depth interviewing (see Chapter 4 for details) or observational methods (see Chapter 6 for details) that, though time-consuming, often result in a deeper or more detailed appreciation of the complicated issues involved (Geertz 1973; Herod 1993).

Talking with, or observing people as long as necessary (or at least as long as possible), to understand their different experiences and perspectives is a good way to develop an in-depth understanding of the positions and issues surrounding any particular research interest. In point of fact, it is perfectly feasible that conducting in-depth interviews with a small number of the '*right*' people will provide significant insights into a research issue. In selecting the right people, however, we rely heavily on early and ongoing secondary research to allow us to approach appropriate key informants to unlock a topic with us. The next matter to be resolved is how to select participants for possible involvement in our research.

How to select participants

Michael Patton's (1990, pp. 182–3) work on **purposeful sampling** is among the more useful summaries on the topic available to researchers. Patton refers to fifteen forms of purposeful sampling, including the following commonly employed strategies. *Extreme* or *deviant case* sampling is designed to help researchers learn from highly unusual cases of the issue of interest, such as outstanding successes/notable failures, top-of-the-class students/dropouts, exotic events, or crises. *Typical case* sampling illustrates or highlights what is considered 'typical', 'normal' or 'average' (such as the early 1990s television documentary *Sylvania Waters*, which depicted life in a 'typical' Australian family). *Maximum variation* sampling documents unique or diverse variations that have emerged in adapting to different conditions and identifies important common patterns that cut across variations. *Snowball* or *chain* sampling identifies cases of interest from people who know other people with relevant cases (such as Kirby's and Hay's [1997] work involving homosexual men in Adelaide). *Criterion* sampling involves picking all cases that meet some criterion, such as all children held back a grade at some time in their schooling. *Opportunistic* sampling requires that the researcher flexibly follows new leads during fieldwork and takes advantage of the unexpected. *Convenience* sampling involves selecting cases or participants on the basis of access (for example, interviewing passers-by in the street). While this final strategy saves time, money and effort, it often provides the lowest level of dependability, and can yield information-poor cases. Much purposeful sampling combines a number of these strategies.

How many participants to select

In both qualitative and quantitative research it is usual that only a subgroup of people or phenomena associated with a case is actually studied. The size of this group is more relevant in quantitative research where representativeness is important. In qualitative research, however, the sample is not intended to be representative since the 'emphasis is usually upon an analysis of meanings in specific contexts' (Robinson 1998, p. 409).

Some of the ways in which the issue of how many respondents to select is different in qualitative as opposed to quantitative research are introduced in the following analogy between a case and an island. Suppose you are looking at a special kind of aerial photograph of an island, so detailed that you can see all of its inhabitants.

> Clearly, if the population of the island were ten thousand instead of ten, enumeration would count for a great deal. . . . But this is because of the investigator's limitations: [s/he] cannot really get to know ten thousand people and the various ways in which each interacts with others. The use of formalist techniques is a second-best approach to this problem because the ideal technique is no longer feasible. Even on this big island, the old technique will count for a great deal, but that is not the main point. The point is that counting and model building and statistical estimation are not the primary methods of scientific research in dealing with human interaction: they are rather crude second-best substitutes for the primary technique, storytelling.[1] (Ward 1972, p.185)

Numbers *do* tell us things about the island, and if what we are interested in is the frequency and geographic distribution of the island's population then we need no more than the photograph. If we are interested in a particular 'story', such as might revolve around an aspect of the cultural geography of the island, for example multiple-use conflict in public places, then we will need more than the photograph to go on.

One way to conduct our investigation is to talk with the island's inhabitants. We could also engage in participant observation or we could consult relevant texts such as submissions to government, letters to the editor of the island's newspaper, or television news stories that might give us an insight into multiple-use conflict in the island's public places. As researchers, however, we are usually resource-limited, both in

1 In the sense meant by Ward, storytelling is not 'yarning'. Research storytelling is more about bricolage. '[A] bricoleur's work … [is] a freely assembled collection of disparate items ordered to form a composite, refunctioned whole' (Reed and Harvey 1992, p. 354). This whole is a story fashioned and told by the researcher, not on the basis of imagination or exaggeration, but on the grounds of purposeful investigation and evidence in 'a joint production' (Mishler 1986, p. 82) with research participants. (For more detail on bricolage see Herod 1993; Jackson Lears 1985; Linstead and Grafton-Small 1990; Weinstein and Weinstein 1991; Nelson et al. 1992; Denzin and Lincoln 1994; Thrift 1996. For something of a deconstruction of bricolage, see Derrida 1976, pp. 138–139; 1978, pp. 278–293.)

terms of funding and time, and we must make decisions about what/who to include and what/who to exclude from our study. It is clear, however, that we remain faced with the issue of how many people to talk with or how many texts to read and so forth. While it may seem disconcertingly imprecise, Patton's (1990, p. 184 and 185) brutally simple advice remains accurate:

> *There are no rules for sample size in qualitative inquiry.* Sample size depends on what you want to know, the purpose of the inquiry, what's at stake, what will be useful, what will have credibility, and what can be done with available time and resources....
> *The validity, meaningfulness, and insights generated from qualitative inquiry have more to do with .... information-richness .... and the observational/analytical capacities of the researcher than with sample size.* (emphasis in original)

In the final analysis, then, it is you as the researcher who must be able to justify matters of case and participant selection to yourself, your supervisor, your interpretive community and reader-users of your work.

ENSURING RIGOUR

It is no frivolous thing to share, interpret and represent others' experiences. We need to take seriously 'the privilege and responsibility of interpretation' (Stake 1995, p. 12). This responsibility to informants and colleagues means that it must be possible for our research to be evaluated. It is important that others using our research have reason to believe that it has been conducted dependably. (An extensive literature on these matters includes works by Baxter and Eyles 1997; Bogdan and Biklen 1992; Ceglowski 1997; Denzin 1979; Dey 1993; Flick 1992; Geertz 1973; Jick 1979; Johnson 1997; Keen and Packwood 1995; Kirk and Miller 1986; Manning 1997; Mays and Pope 1997.)

Ensuring rigour in qualitative research (Box 3.3) means establishing the *trustworthiness* of our work (Baxter and Eyles 1999a, 1999b; Bailey et al. 1999a, 1999b). Research can be construed as a kind of 'hermeneutic circle' starting from our interpretive community, and involving our research participant community and ourselves, before returning to our interpretive community for assessment (Reason and Rowan 1981; Burawoy et al. 1991; Jacobs 1999). This circle is a key part of ensuring rigour in qualitative research; our participant and interpretive communities check our work for credibility and good practice. In other words, trust in our work is not assumed but has to be earned.

Box 3.3: Considerations in research design, stage 3: the hermeneutic research circle and checks for rigour

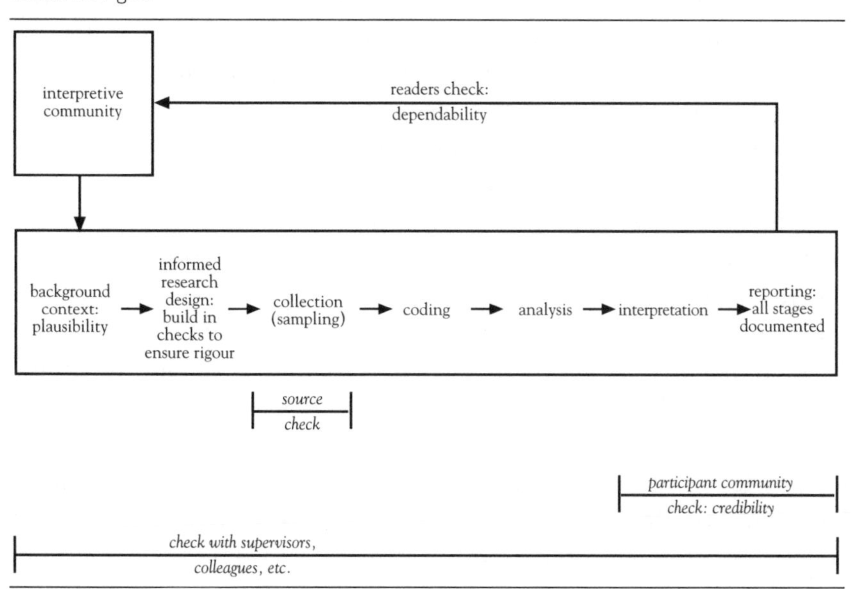

There are two particular steps that need to be followed to ensure and argue the rigour of our research for our interpretive communities. First, strategies for ensuring trustworthiness need to be formulated in the early stages of research design and applied at various stages in the research process (Lincoln and Guba 1985; Baxter and Eyles 1999a, 1997). These should include appropriate checking procedures in which our work is opened up to the scrutiny of interpretive and participant communities. Second, we need to document each stage of our research carefully so that we might report our work to our interpretive community for checking; 'we should focus on producing analyses that are as open to scrutiny as possible' (Fielding 1999, p. 526).

Rigour is a matter that needs to be considered from the outset of our research, underpinning the early stages of research design. It is important to incorporate appropriate *checking* procedures into our research process. These procedures are outlined in Denzin (1978) and Baxter and Eyles (1997) as the four major types of **triangulation**: multiple sources, methods, investigators and theories. For example, as we move through various research stages, we might check: (a) our sources against others (re-search); (b) our process and interpretations with our supervisors and/or colleagues; and (c) our text with our research participant community to enhance the credibility of our research (although this last check can be problematic if that community has considerable power, such as might be the case with a multi-national corporation whose managers refuse us permission to publish work related to findings derived from the corporation). Reason and Rowan (1981) elaborate on some of these matters.

Examining the research stages in Box 3.3—which often overlap as they become a whole research composition—we also need to document our work fully: how we came to be interested in the research, why we chose to do it, and for what purpose. We may declare our own philosophical, theoretical and political dispositions, and we will almost certainly review literatures dealing with both the general area of our research and the research methods we intend to use. This elaboration of context permits us to establish the plausibility of our research by demonstrating that we embarked on our work adequately informed by relevant literatures and for intellectually and ethically justifiable reasons. We will most likely have checked the plausibility of our research with supervisors and/or colleagues before embarking on detailed research design. At the final stage of reporting research we can also attempt to acknowledge limits to the transferability of our research due to particularities of the research topic, the research methods used and the researcher. In this way, we confirm that the methods we use and the interpretations we invoke influence our research outcome. Thus, it is vital that we document all stages of our research process. Such documentation allows members of our interpretive and participant communities to check all of these stages so our work might be considered dependable.

FINAL COMMENTS

We began this chapter by suggesting that consideration of research design and rigour is essential to the conduct of dependable qualitative inquiry. We have addressed issues of case selection and participant selection, and outlined some reasons to be concerned with rigour, as well as some means by which rigour might be achieved.

Most research is undertaken to be shared with others. We therefore need to ensure that our research can stand up to the critical scrutiny of our interpretive and participant communities. The work presented in this chapter provides some of the conceptual and practical tools by which this outcome of sharing plausible, credible and dependable work can be achieved.

KEY TERMS

case
disconfirming case
interpretive community
participant
purposeful sampling
rigour
triangulation

REVIEW QUESTIONS

1. Why is rigour important in qualitative research?
2. What is an 'interpretive community'?
3. What is meant by the phrase 'participant community'?
4. What are some ways we might check our research to establish its dependability to members of our interpretive community?

SUGGESTED READING

Baxter, J. and Eyles, J. 1997, 'Evaluating qualitative research in social geography: establishing 'rigour' in interview analysis', *Transactions of the Institute of British Geographers*, vol. 22, pp. 505–525.

Jacobs, J. 1999, 'The labour of cultural geography', in *Australian Cultural Geographies*, ed. E. Stratford, Oxford University Press, Melbourne.

Patton, M.Q. 1990, *Qualitative Evaluation and Research Methods*, 2nd edn, Sage, Beverly Hills.

Platt, J. 1988, 'What can case studies do?', *Studies in Qualitative Methodology*, vol. 1, pp. 1–23.

Sayer, A. 1992, *Method in Social Science: A Realist Approach*, 2nd edn, Routledge, New York.

Interviewing

Kevin Dunn

Contents

CHAPTER OVERVIEW

This chapter provides advice on interview design, practice, transcription, data analysis and presentation. I describe the characteristics of each of the three major forms of interviewing and critically assess what I see as the relative strengths and weaknesses of each. Applications of interviewing are outlined by referring to examples from economic, social and environmental geography.

INTERVIEWING IN GEOGRAPHY

Interviewing in geography is so much more than 'having a chat'. Successful interviewing requires careful planning and detailed preparation. A taped hour-long interview will require days of preliminary background work and question formulation, as well as the diplomacy of contacting **informants** and negotiating 'research deals'. A sixty-minute interview will require at least four hours of transcription if you are a fast typist and verification of the record of interview could stretch out over a couple of weeks. After all that, you have still to analyse the interview material. These are involved and time-consuming activities. Is it all worth it? In this chapter I outline some of the benefits of interviewing, and provide a range of tips for good interviewing practice.

An interview can be defined as 'a face-to-face verbal interchange in which one person, the interviewer, attempts to elicit information or expressions of opinion or belief from another person or persons' (Maccoby and Maccoby 1954, p. 499). An interview is a data gathering method in which there is a spoken exchange of information. It is a method which requires some form of direct access to the person being interviewed.

Types of interviewing

There are three major forms of interviewing: structured, unstructured and semi-structured. These three forms can be placed along a continuum with the **structured interview** at one end and the **unstructured interview** at the other (see also Chapter 1, Box 1.1). Structured interviews follow a predetermined and standardised list of questions. The questions are always asked in the same way and in the same order. At the other end of the interviewing continuum are unstructured forms of interviewing such as oral histories. The conversation in these interviews is actually directed by the informant rather than by set questions. In the middle of this continuum are **semi-structured interviews**. This form of interviewing has some degree of predetermined order but still ensures flexibility in the way issues are addressed by the informant. Different forms of interview have varying strengths and weaknesses which should be clear to you by the end of this chapter.

Strengths of interviewing

Research interviews are used for four main reasons (see also Krueger 1994; Minichiello et al. 1995, pp. 70–4):

1. to fill a *gap in knowledge* which other methods, such as observation or the use of census data, are unable to bridge efficaciously;
2. to investigate *complex behaviours and motivations*;
3. to collect a *diversity of opinion and experiences*. Interviews provide insights into the differing opinions or debates within a group, but they can also reveal consensus on some issues; and
4. when a method is required that shows *respect for* and *empowers* those people who provide the data. In an interview the informant's view of the world should be valued and treated with respect. The interview may also give the informant cause to reflect on their experiences and the opportunity to find out more about the research project than if they were simply being observed or if they were completing a questionnaire.

Interviews are an excellent method of gaining access to information about events, opinions and experiences. While opinions and experiences vary enormously between people of different class, ethnicity, age and sexuality, interviews have allowed me to understand how meanings differ among people. Geographers who use interviewing should be careful to resist claims that they have discovered the *truth* about a series of events, or that they have distilled *the* public opinion (Goss and Leinbach 1996, p. 116; Kong 1998, p. 80). Interviews are used to gain *insight* into *different* opinions. They can also be used to counter the claims of those who have presumed to have discovered *the* public opinion. This can be done by seeking out the opinions of different groups, often marginalised or subaltern groups, whose opinions are rarely heard.

Most of the questions posed in an interview allow for an open response, as opposed to a closed set of response options such as Yes or No. In this way, each informant can advise you as researcher about events or opinions in their own words. One of the major strengths of interviewing is that it allows you to discover what is relevant to the informant.

Due to the face-to-face verbal interchange used in interviewing the informant can tell you if a question is misplaced (Box 4.1). Furthermore, your own opinions and tentative conclusions can be checked, verified and scrutinised. This may disclose significant misunderstandings on your part or issues which you had not previously identified (Schoenberger 1991, p. 187).

Box 4.1: Asking the wrong question: a tale from Cabramatta

On 23 June 1990, I began my first formal research interview.The informant was a senior office-bearer from one of the Indo-Chinese cultural associations in New South Wales. My research interest was in the social origins of the concentration of Indo-Chinese-Australians around Cabramatta and the experiences of these migrants and their families.The political context of the time was still heavy with the racial and anti-Asian overtones of the 1988 'immigration debate' in which mainstream politicians and academics such as John Howard and Geoffrey Blainey had expressed concern about 'Asian' immigration and settlement patterns in Australian cities. Specifically, Vietnamese migrants were accused of congregating in places like Cabramatta, Richmond and Springvale, and of purposefully doing so in order to avoid participating with the rest of Australia. I was a somewhat naïve and colonialist investigator, who saw his role as 'valiant protector' of an ethnic minority. I therefore hypothesised that Vietnamese-Australians did not congregate voluntarily, but that they were forcibly segregated by the economic and social constraints of discrimination in housing and labour markets. Indeed, the geographic literature supported my assertion at the time.

But back to my first interview. My opening question was: 'Please explain the ways in which discrimination has forced you, and members of the community you represent, to reside in this area?' The answer: 'I wouldn't live anywhere else'. This informant, and most subsequent informants, described the great benefits and pleasures of living in Cabramatta. They also explained how residing in Cabramatta had eased and assisted their expanding participation in Australian life (Dunn 1993). I had asked the wrong questions and had been told so by my informants. I decided to focus on the advantages and pleasures which residence in 'Cab' brought to Indo-Chinese-Australians. The face-to-face nature of the exchange, and the informed subject, makes interviews a remarkable method. The participants can tell the researcher: 'you're wrong'!

INTERVIEW DESIGN

It is not possible to formulate a strict guide to good practice for every interview context. Every interview and every research issue demands its own preparation and practice. However, researchers should heed certain procedures. Much of the rest of this chapter focuses on strategies for enhancing the credibility of data collected using rigorous interview practice. In the next section we look at the organisation of **interview schedules** and the formulation of questions.

The interview schedule or guide

Even the most competent researcher needs to be reminded during the interview of the issues or events they had intended to discuss. You cannot be expected to recall all of the specific questions or issues you wish to address and you will benefit from some written reminder of the intended scope of the interview. These reminders can take the form of interview schedules or **interview guides**.

An interview guide or ***aide mémoire*** (Burgess 1982), is a list of general issues you want to cover in an interview. Guides are usually associated with semi-structured forms of interviewing. The guide may be a simple list of key words or concepts intended to remind you of discussion topics. The topics initially listed in a guide are often drawn from existing literature on an issue. The identification of key concepts and the isolation of themes is a preliminary part of any research project (see Babbie 1992, pp. 88–164).

One of the advantages of the interview guide is its flexibility. As the interviewer, you may allow the conversation to follow as 'natural' a direction as possible, but you will have to re-direct the discussion to cover issues that may still be outstanding. Questions can be crafted *in situ*, drawing on themes already broached and from the tone of the discussion. The major disadvantage of using an interview guide is that you must formulate coherent questions 'on the spot'. This requires good communication skills and a great deal of confidence. Any loss of confidence or concentration may lead to inarticulate and ambiguous wording of questions. Accordingly, a guide is inadvisable for first-time interviewers. It is more appropriate for very skilled interviewers and for particular forms of interview, such as **oral history**.

An *interview schedule* is used in structured, and sometimes semi-structured, forms of interviewing. They are also called question schedules or 'question routes' (Krueger 1994). An interview schedule is a list of carefully worded questions (see Box 4.2).

Box 4.2: Formulating good interview questions

- Use easily understood language that is appropriate to your informant.
- Use non-offensive language.
- Use words with commonly and uniformly accepted meanings.
- Avoid ambiguity.
- Phrase each question carefully.
- Avoid leading questions as far as possible (i.e., questions that encourage a particular response).

I have found that a half-hour interview will usually cover between six and eight primary questions. Under each of these central questions I nest at least two detailed questions or prompts. In some research it may be necessary to ask each question in the same way and in the same order to each informant. In others, you might ask questions at whatever stage of the interview seems appropriate. The benefits of using interview schedules mirror the disadvantages of interview guides. They provide greater confidence to the researcher in the enunciation of their questions and allow better comparisons between informant answers. However, questions which are prepared before the interview and then read out formally may sound insincere, stilted and out of place.

A mix of carefully worded questions and topic areas capitalises on the strengths of both guides and schedules. Indeed, a fully worded question can be placed in a guide and yet be used as a topic area. The predetermined wording can be kept as a 'fall-back' in case you find yourself unable to articulate a question 'on the spot'. I find it useful to begin an interview with a prepared question. There is little that is more damaging to one's confidence than the informant saying 'what do you mean by that?' or 'I don't know what you mean' in response to your first question.

Interview design should be *dynamic* throughout the research (Tremblay 1982, pp. 99 and 104). As a research project progresses you can make changes to the order and wording of questions/topics as new information and experiences are fed back into the research design. Some issues may be revealed as unimportant, silly, or offensive after the initial interview. These can be dropped from subsequent interviews. The interview schedule or guide must also garner information in a way that is appropriate to each informant.

While the primary purpose of interview schedules and guides is to jog your memory and to ensure that all issues are covered as appropriately as possible, it is also useful to provide informants with a copy of the questions or issues before the interview to prompt thought on the matters to be discussed. Interview guides and schedules are also useful note-taking sheets (Webb 1982, p. 195).

TYPES OF QUESTIONS

Interviews utilise **primary** (or original) **questions** and **secondary questions**. Primary questions are opening questions used to initiate discussion on a new theme or topic. Secondary questions are **prompts** which encourage the informant to follow-up or expand on an issue already discussed. An interview schedule, and even an interview guide, can have a mix of types of original questions, including descriptive questions, storytelling prompts, structural questions, contrast and opinion questions, and devil's advocate propositions (see Box 4.3). Since different types of primary questions produce very different sorts of responses, a good interview schedule will generally comprise a mix of question types.

Box 4.3: Primary question types

Type of question	*Example*	*Type of data and benefits*
Descriptive	*What is the full name of your organisation?*	Details on events, places, people and experiences.
(knowledge)	*What is your role within the organisation?*	Easy-to-answer opening questions.
	How many brothers or sisters do you have?	
Storytelling	*Can you tell me about the formation and history of this organisation and your involvement in it?*	Identifies a series of players, an ordering of events or causative links. Encourages sustained input from the informant.
Opinion	*Is New Zealand society sexist?*	Impressions, feelings, assertions and guesses.
	What do you consider to be the appropriate size for a functional family?	
Structural	*How do you think you came to hold that opinion?*	Taps into people's ideology and assumptions.

	What do you think the average family size is for people like yourself?	Encourages reflection on how events and experiences may have influenced opinions and perspectives.
Contrast (hypothetical)	*Would your career opportunities have been different if you were a man? or if you grew up in a poorer suburb?*	Comparison of experience by place, time, gender and so forth. Encourages reflection on (dis)advantage.
Devil's advocate	*Many of your own colleagues are privately voicing concern about your policy. Are you about to ....?* See also Box 4.4.	Controversial/sensitive issues broached (opinions of political opponents) without associating the researcher with that opinion.

Box 4.4: Asking the tough questions without sounding tough

My work with Indo-Chinese communities in Cabramatta occurred within a political context in which Vietnamese-Australians were being publicly harangued by academics and politicians (Dunn 1993). I felt it was important to get the informants to respond to the views of their critics. My interview schedule had the following two devil's advocate propositions. I also used a preamble to dissociate myself from the statements.

> *In my research so far, I have come across two general explanations for Vietnamese residential concentration. I would like you to comment on two separate statements, that to me, represent these two explanations:*
> *First: that Vietnamese people have concentrated here because they don't want to participate with the wider society.*
> *Second: that the Vietnamese are segregated into particular residential areas through social, economic and political forces imposed upon them by the wider society.*

The aim was to gather peoples' responses to both statements. In most cases, informants were critical of both views. Some informants took my question as a request for them to select the explanation which they thought was the most appropriate. Those people selected the second statement. Others were in no doubt that I disagreed with both views. Either way, devil's advocate propositions are often leading. My political views were noticeable in the question preamble and wording, as well as in the preliminary discussions held to arrange the interviews. It is fairer that the researcher's motives and political orientation are obvious to the informant rather than hidden until after the research is published (see Chapters 2, 3 and 9 of this volume).

On an interview guide or schedule you might have a list of secondary questions or prompts (Box 4.5). There is a number of different types of prompts, ranging from formal secondary questions to nudging-type comments which encourage the informant to continue speaking (Whyte 1982, p. 112). Sometimes prompts are listed in the interview guide or schedule, but often they are deployed, when appropriate, without prior planning.

Box 4.5: Types of prompt

Prompt type	*Example*	*Type of data and benefits*
Formal secondary question	Primary Q: *What social benefits do you derive from residing in an area of ethnic concentration?*	Extends the scope or depth of treatment on an issue.
	Secondary Q: *What about informal child-care?*	Can also help explain/rephrase a misunderstood primary question.
Clarification	*What do you mean by that?*	Used when an answer is vague or incomplete.
Nudging	*And how did that make you feel?* Repeat an informant's last statement.	Used to continue a line of conversation.
Summary (categorising)	*So let me get this straight: your view, as just outlined to me, is that people should not watch shows like 'Neighbours'?*	Outline in-progress findings for verification. Elicit succinct statements (for example, 'quotable quotes').
Receptive cues	Audible: *Yes, I see. Uh-huh.* Non-audible: nodding and smiling.	Provides receptive cues, encourages an informant to continue speaking.

Ordering questions and topics

It is important to consider carefully the order of questions or topics in an interview guide or schedule. Minichiello et al. (1995, p. 84) advise that the most important consideration in question order is preserving **rapport** between you and your informant. This requires that discomfort for the informant be minimised. There are two broad sorts of advice regarding the order in which issues should be addressed: the 'funnel' and the 'pyramid structure' approaches.

Funnelling involves an initial focus on general issues, followed by a gradual movement towards personal matters and issues specific to the informant. This strategy allows for conversational development towards more sensitive issues: 'The assumption made in using this strategy is that informants and interviewers would find it uncomfortable to start talking directly about an issue which may be personally threatening or uncomfortable to think about' (Minichiello et al. 1995, p. 84).

As an ordering strategy funnelling draws on long-held advice in interviewing to keep sensitive questions until the end (Sudman and Bradburn 1982, pp. 73 and 78). The advantage of this strategy is that the interview begins in a relaxed and non-threatening manner. Rapport between informant and interviewer can develop and the chance that the informant will discontinue the interview is reduced. Even if an informant ends an interview at a sensitive point, you will nonetheless have gathered some data. Funnelling might usefully be employed, for example, when investigating an informant's experiences of oppression. The interview might begin with a discussion of the general problem of homophobia or racism: how widespread it is; how it varies from place to place; and how legal and institutional responses to those forms of oppression have emerged. Having broached these general or macro-level aspects of oppression the interview might then turn to the particular experiences of the informant.

In the **pyramid interviewing strategy** abstract and general questions are asked at the end. The interview starts with easy-to-answer questions about an informant's duties or responsibilities, or their involvement in an issue. This allows the informant to become accustomed to the interview, interviewer and topics before they are asked questions which require deeper reflection. For example, to gather views on changes to urban governance you might find it necessary to first ask an informant from an urban planning agency to outline their roles and duties. Following that, you might ask your informant to outline the actions of their own agency and how those actions may have changed in recent times. Once the 'doings and goings-on' have been outlined it may then make sense to ask the informant why agency actions and roles have changed, whether that change has been resisted, and how they view the transformation of urban governance.

A final question-ordering option is to use a hybrid of funnel and pyramid structures. The interview might start with simple-to-answer, non-threatening questions, then move to more abstract and reflective aspects, before gradually progressing towards sensitive issues. This sort of structure may offer the benefits of both funnel and pyramid ordering.

When thinking about question and topic ordering, it can be helpful to have key informants comment on the interview guide or schedule (Kearns 1991, p. 2). Key informants are often initial or primary contacts in a project. They are usually the first informants and they often possess the expertise to liaise between the researcher and the communities being researched. Key informant review can be a useful litmus test of interview design, since these representatives are 'culturally qualified'. They have empathy with the study population, and can be comprehensively briefed on the goals and background of the research (Tremblay 1982, pp. 98–100).

STRUCTURED INTERVIEWING

A structured interview uses an interview schedule which typically comprises a list of carefully worded and ordered questions (see Boxes 4.3 and 4.5 and the earlier discussion on ordering questions and topics). Each respondent or informant is asked exactly the same questions in exactly the same order. The interview process is *question focused.*

It is a wise idea to pre-test a structured interview schedule on a subset (say 3–10) of the group of people you plan to interview for your study to ensure that your questions are not ambiguous, offensive or difficult to understand. Though helpful, 'pre-testing' is of less importance in semi-structured and unstructured interviews, where ambiguities (but not offensive questions!) can be clarified by the interviewer.

Structured interviews have been used with great effect throughout geography's sub-disciplines, including economic geography (Box 4.6).

Box 4.6: Interviewing—an economic geography application

In 1991 Schoenberger argued that most economic geography approaches had been on the outside looking in, deducing strategic behaviour from its locational effects rather than investigating it directly (1991, p. 182). One of the assumptions challenged by the use of structured interviews was that the location of firms was strongly associated with the location preferences of the industries in question. Using structured interviews with managers, Schoenberger was able to show that the location of foreign chemical firms in the United States was as much, if not more, related to historical and strategic contingencies than to contemporary location preferences.

For example one of Schoenberger's case studies was a German-owned chemical firm. Her interviews revealed that the firm's Board of Directors had decided on a major expansion in the US market. But the Board had been split between establishing a 'greenfields' site which would be purpose-built to company needs, and acquiring an already established chemical plant which would hasten their expanded presence in the market. Plans to establish a greenfield facility were foiled by organised community opposition. The directors who argued for an acquisition then gained the upper hand, and at about the same time a US chemical firm came up for sale. For many decades German chemical firms had agreed among each other to specialise in certain parts of the chemical sector. These agreements were about to end, and Schoenberger's case-study firm were keen to expand horizontally into another speciality. The US firm which came up for sale happened to specialise in that area. A host of historical and strategic events had combined to produce a particular location result.

The historical and strategic contingencies which accounted for the location of the German chemical company were revealed through structured interviews. Their location was in fact quite at odds with the apparent preferences of the firm and reveals nothing about the firm's location preferences (Schoenberger 1991, p. 185). The US chemical sector has a high level of foreign ownership, and most of it was established through acquisition. Replicated interviews with managers and directors across the chemical sector revealed the prevalence of location choices being determined by historical and strategic contingency. Interviewing was therefore an essential method for unravelling the location determinants of chemical plants in the US.

SEMI-STRUCTURED INTERVIEWING

Semi-structured interviews employ an interview guide. The questions asked in the interview are *content focused* and deal with the issues or areas judged by the researcher to be relevant to the research question. Alternatively, an interview schedule might be prepared with fully worded questions for a semi-structured interview, but you would not be restricted to deploying those questions. The semi-structured interview is organised around ordered but flexible questioning. In semi-structured forms of interview the role of the researcher (interviewer or facilitator) is recognised as being more interventionist than in unstructured interviews. This requires that you redirect the conversation if it has moved too far from the research topics (see also Chapter 5).

UNSTRUCTURED INTERVIEWING

Various forms of unstructured interviewing exist. These include oral history, **life history**, and some types of group interviewing and in-depth interviewing. Unstructured interviewing focuses on personal perceptions and personal histories. Rather than being question focused like a structured interview, or content focused as in a semi-structured format, the unstructured interview is *informant focused*. Life history and oral history interviews seek personal accounts of significant events and perceptions, as determined by the informants, and in their own words. Each unstructured interview is unique. The questions you ask are almost entirely determined by the informant's responses. These interviews approximate normal conversational interaction and give the informant some scope to direct the interview. Nonetheless, an unstructured interview requires as much, if not more, preparation than its structured counterpart. You must spend time sitting in musty archive rooms or perched in front of dimly lit microfiche machines gaining a solid understanding of past events, people and places related to the interview.

We will explore two types of unstructured interviewing below: oral history and life history. These are often confused, but they have very different aims and backgrounds.

Oral history

The rise of oral history as a popular research method was associated with an increasing sense of dissatisfaction with histories reviewing the actions and opinions of the rich and powerful. Robertson writes, 'One of the most important uses of oral history is to record the perspectives of disadvantaged people who traditionally have been either ignored or misrepresented in conventional historical records' (1993, p. 3).

Since the 1960s, oral history in Australia has been seen as a way of making a 'people's history' more prominent (Douglas, Roberts and Thompson 1988). Oral history provides perspectives on past events from those people who were not recorded in the newspapers or chronicles of the day, or who were unable to write and so did not keep diaries or correspond with other people. We may 'find out about' events and places which had been kept out of the news, or which had been deemed of no consequence to the rich and powerful (Box 4.7). It also gives researchers access to people who may have been purposefully silenced or prevented from 'speaking out'. This type of interview is not however confined to eliciting the views of powerless groups. In general, oral histories allow researchers to expand the subject matter of historical geography, to gain alternative perspectives on the 'main events', and to propose hitherto unmentioned events as significant. They contribute to a diverse historical record.

Box 4.7: Oral environmental histories

Oral history interviews can collect data about environmental history. This type of interviewing helps produce a more comprehensive picture of the cause and process of environmental change than is available through physical methods of enquiry. Data collected might include peoples' memories of changes in local land use, biodiversity, hydrology and climate.

Lane (1997) used oral history interviews to reveal changes in watercourses, weeds and climate in the Tumut Region high country of the Australian Alps. Interviews were conducted with five main informants, firstly in their homes, and then while driving and walking through the countryside where they had resided. The informants told of the waterholes and deep parts of creeks where they would fish and swim, and where they and their children had learnt to swim. One informant commented that one of the creeks used to be almost a river ... and now you could step over it (Lane 1997, p. 197). The same informant noted the change in colour and quality of the water. Lane's informants described how the water level and quality had steadily degraded since pine plantations had been planted in the 1960s. This description was consistent with 'scientific' understandings of the impact of pine plantations in which there is an ever decreasing level of run-off as the pines grow.

Such specific observations from local residents may often be the only detailed evidence on watercourse change available. Oral history can fill gaps in the 'scientific record' or it can be used to complement data gathered using physical or quantitative methods. More importantly, with the use of oral history environmental change can be set in a human context and related to the history of people who lived in the region (Lane 1997, p. 204).

Life history

Life histories are interviews in which the researcher attempts to elicit information about the experiences and development of an informant's entire life. Unlike oral histories, the research focus is not on the informant's role in or perspective on key historical events or times, but rather on how the informant's life has unfolded. Some of the most famous uses of life history have been projects aimed at revealing the ways people learn and negotiate their sexuality or gender (Bogdan 1974; Connell 1991). A research project using life histories might typically involve five or six carefully selected people who are prepared to reflect on and discuss their life experiences and development. The questions asked in the interview are triggers intended to get the informant talking. Triggers include asking people to comment on important events, turning points and transitions in their lives. Turning points include entry to the waged labour market, migration, and various rites of passage. They might have been identified in earlier interview sessions, in which the informant quickly recounted the main events in his/her life (Connell 1991, p. 144; Findlay and Li 1997, pp. 37–8). The researcher can then reflect on the events described, and in a subsequent interview ask the informant to expand upon their role in, feelings about or reactions to such occurrences (Box 4.8).

Box 4.8: Life history in migration studies

Life history has been used by geographers to analyse migration. It has been used to uncover the individual decisions which underpin migration behaviour (Findlay and Li 1997), and to reveal the very specific cultural impacts of this turning point upon individual migrants (Pulvirenti 1997). These geographies have begun with the premise that migrants are active social agents capable of employing their knowledge of structures to achieve their own goals and, by their collective actions, capable of reproducing and transforming structures (Findlay and Li 1997, p. 34).

Findlay and Li were convinced that migration decisions can only be properly understood in the context of an individual's value system as developed over their life course. In order to analyse the migration motivations, intentions and experiences of Hong Kong emigrants they undertook life histories, or what they called auto-biographies, with 40 professionals who had emigrated to Canada and the United Kingdom. Two interviews were carried out with each informant. The first interview was used to identify the key turning points or factors which the informants thought important to their migration decision. The informants made reference to a fairly restricted set of migration catalysts such as career advancement and educational opportunities. In the second interview sessions the informants began to identify a much more diverse array of events, memories and influences which helped explain their move overseas. These additional factors included: advertising pictures of countries like Canada and the United Kingdom which had been seen over many years, popular sayings about living overseas and its assumed social status, photos sent back by friends already studying or working overseas, as well as a pursuit of freedoms (from family, peers, the state). Family histories and circumstances, as well as Hong Kong's cultural and economic structures, were found to be as important to migration decisions as are individual motivations of socio-economic betterment.

The life history interview revealed that an individual's migration decision is situated in his or her life history, rather than just in the moment when the decision is made (Findlay and Li 1997, p. 38).

The great advantage of life history is the ability to gain insights into how individuals interact with society as their lives change. Connell (1991) used life history interviews with Australian men to 'tap into' the ways masculinity is learnt and practised. Life history interviews provide data on concrete experiences and events, but they also yield evidence on how people move through transitional periods and how they interact with institutions. In this way the life history interview can collect data on both individuals (their actions, feelings) and social structures (collectives, institutions, milieux). This attention to the interplay between structure and agency convinced Connell (1991, pp. 143–4) that life history interviews could be used 'to understand the construction of gender as a project in time'. Most social scientists, including human geographers, have a keen interest in social structures and social change. This makes the life history a useful research method in geography (Box 4.8).

INTERVIEWING PRACTICE

Rapport with another person is basically a matter of understanding their model of the world and communicating your understanding symmetrically. This can be done effectively by matching the perceptual language, the images of the world, the speech patterns, pitch, tone, speed, the overall posture and the breathing patterns of the informant. (Minichiello et al. 1995, p. 80)

Achieving and maintaining rapport, or a productive interpersonal climate, can be critical to the success of an interview. Rapport is particularly important if you need to have repeat sessions with an informant. Even the first steps of arranging an interview are significant, including the initial contact by telephone and other preliminaries that might occur before the first interview. Interviews in which both the interviewer and informant feel at ease usually generate more insightful and more valid data than might otherwise be the case. In the following paragraphs, I outline a set of tips which can help you enhance rapport before, during and while closing an interview.

Contact

Informants are usually chosen purposefully on the basis of the issues and themes which have emerged from a review of previous literature or from other background work (see Chapter 3). This involves choosing people who can communicate aspects of their experiences and ideas relevant to the phenomena under investigation (Minichiello et al. 1995, p. 168). Decisions about the selection of informants also depend on your ability to gain access to people. Once you have identified a potential informant you must then negotiate permission for the interview. This means getting the consent of the informants themselves and, in some circumstances, it will also involve gaining the sanction of 'gatekeepers' like employers, parents or teachers. This might occur for example if you wanted to interview school children, prisoners, or employees in their workplace.

Your first contact with an informant will often be by telephone, or by some form of correspondence. In this preliminary phase you should do at least four things (Robertson 1994, p. 9):

1 Introduce yourself and establish your bona fides. For example: 'My name is Chuck McGutzup and I am an honours student from Puke State University'.
2 Make it clear how you came to get the informant's name and telephone number or address. If you do not explain this people may well be suspicious and are likely to ask how you got their name. If you are asked this question, rapport between you and your informant has already been compromised.
3 Outline why you would like to conduct the interview with this informant in particular. Indicate the significance of the research and explain why the informant's views and experiences are valued. For instance, you may believe they have important things to say, that they have been key players in an issue, or that they have experienced something specific which others have not. On the whole, I have found that most people are flattered to be asked for an interview, although they are often nervous or hesitant about the procedure itself.
4 Indicate how long the interview and any follow-up is likely to take.

Making an informant feel relaxed involves dealing with all of the issues mentioned above, and in addition spelling out the mechanics of the interview and negotiating elements of the interview process. All of these can be outlined in a 'Letter of Introduction' which may be sent to an informant once they have agreed to an interview, or while agreement is still being negotiated. This formal communication should be under the letterhead of your organisation (for example, your university), and should spell out your bona fides, the topic of the research, the manner in which the interview will be conducted, and any rules or boundaries regarding confidentiality. You must, of course, seek permission from your supervisor to use the letterhead of an organisation such as a university. In the absence of a letter of introduction informants should be made aware of their rights during the interview. This is sometimes referred to as brokering a 'research deal' or a 'research bargain'. The research deal may be agreed to over the telephone, or just before an interview begins. The deal can be set out in written form. (See Box 4.9 for some of the rights of informants which can be established. Chapter 2 of this volume includes material relevant to the ethics of interviewing.) These preliminary discussions are important to the success of an interview. Indeed they set the tone of the relationship between interviewer and informant.

Box 4.9: Codifying the rights of informants

In their research on the Carrington community in Newcastle, New South Wales, Winchester, Dunn and McGuirk (1997) decided to codify informants' rights in the oral histories and semi-structured interviews which were to be conducted. They included the following list of informants' rights on University letterhead, and gave a copy to each of the informants:

- Permission to tape the interview must be given in advance.
- All transcribed material will be anonymous.
- Tapes and transcripts will be made available to those informants who request them.
- Informants have the right to change an answer.
- Informants can contact us at any time in the future to alter or delete any statements made.
- Informants can discontinue the interview at any stage.
- Informants can request that the tape-recorder be paused at any stage during the interview.

To this list, one might make additional statements (for example, that informants could expect information about the ways in which their contributions to the research would be used). A codification of rights was deemed necessary for two reasons. First, it was done to empower the informants and assure them that they could pause or terminate the interview process whenever they deemed it necessary to do so. Second, the researchers had employed an articulate local resident to conduct the interviews and so it was important that the interviewer was also constantly reminded of the informants' rights.

The interview relation

The relationship established between interviewer and informant is often critical to the collection of opinions and insights. If you and your informant are at ease with each other then the informant is likely to be communicative. However, there are competing views on the nature of the interviewer-informant relationship. On the one hand there is an insistence on 'professional interviewing' and on the other there is 'creative' or empathetic interviewing. Goode and Hatt (in Oakley 1981, pp. 309–10) warn that interviewers should remain detached and aloof from their informants: 'the interviewer cannot merely lose himself [*sic*] in being friendly. He must introduce himself as though beginning a conversation, but from the beginning the additional element of respect, of professional competence, should be maintained ... He is a professional in this situation, and he must demand and obtain respect for the task he is trying to perform' (Goode and Hatt, in Oakley 1981, p. 191).

A very different, indeed opposite, sort of relationship has been proposed by Oakley (1981) and Douglas (1985). In their view, a researcher who remains aloof would undermine the development of an intimate and non-threatening relationship (Oakley 1981, p. 310). Rather than demanding respect from the informant, Douglas' model of 'creative interviewing' insists that each informant must be treated as a 'Goddess' of information and insight. Douglas recommends that researchers humble themselves before the Goddess. The creative or empathetic model of interviewing thus advocates a very different sort of relationship between the informant and interviewer than that recommended for 'professional' interview relations. Overall, there is a range of interview practice that lies between the poles of 'professional' or 'creative' interview relationships.

Decisions about the interview relationship will vary according to the characteristics of both the informant and the interviewer. The cultural nuances of a study group will at times necessitate variations in the intended interaction. However, it is wise to remember that despite any empathy or relationships which are established, the interview is still a formal process of data gathering for research. Furthermore, there is usually a complex and uneven power relationship involved in which information, and the power to deploy that information, flows mostly one way: from the informant to the interviewer (see Chapter 2).

Rapport may increase the level of understanding you have about the informant and what they are saying. There are a number of strategies for enhancing rapport. The first is through the use of respectful preliminary work. The second involves the use of a *warm-up* period just before an interview commences. Douglas (1985, p. 79) advises that rather than getting 'right down to business' it is better to engage in some 'small talk and chit-chat (which) are vital first steps'. This warm-up discussion with an informant could be a chat about the weather, matters of shared personal interest, or 'catching-up' talk. In their surveys and interviews of Vietnamese-Australians in Melbourne, Gardner, Neville and Snell (1983, p. 131) found that 'The success of an interview (when measured by the degree of relaxation of all those present and the ease of conversation) generally depended on the amount of "warm-up" (chit-chat, introductions, etc.)'.

My own warm-up techniques have included giving the informant an overview of the questions I plan to ask, presenting relevant diagrams or maps, as well as discussing historical documents (see also Tremblay 1982, pp. 99 and 103). Maps, diagrams, tables of statistics and other documents can also be used as references or stimuli throughout an interview. If an informant offers you food or drink before an interview it would be courteous to accept them.

You should also have acquainted yourself with the cultural context of the informants before the interview. As Robin Kearns pointed out, 'If we are to engage someone in conversation and sustain the interaction, we need to use the right words. Without the right words our speech is empty. Language matters' (Kearns 1991, p. 2). For instance, you must be able to recognise the jargon or slang and frequently used acronyms of institutions or corporations as well as the language of particular professions or cultural groups.

Listening strategies can improve rapport and the productivity of the interview. Your role as interviewer is not passive, but requires constant focus on the information being divulged by informants, and the use of cues and responses to encourage them. Your role as an active participant in the interview extends well beyond simply asking predetermined questions or broaching predetermined topics. You must maintain an active focus on the conversation. This will help prevent lapses of concentration. You must also avoid 'mental wandering', otherwise you may miss unexpected leads. Moreover, it is irritating to the informant, and a threat to rapport, if you ask a question they have already answered (Robertson 1994, p. 44).

Adelman (1981) advises researchers to maintain a **critical inner dialogue** during an interview. This requires that you constantly analyse what is being said and simultaneously formulate the next question or prompt. You should be asking yourself whether you understand what the informant is saying. Do not let something slide by that you do not understand with the expectation that you will be able to make sense of it afterwards. Minichiello et al. (1995, p. 103) provide a demonstration of how critical inner dialogue might occur: 'What is the informant saying that I can use? Have I fully understood what this person is saying? Maybe, maybe not. I had better use a probe. Oh, yes I did understand. Now I can go on with a follow-up question'.

Strategies to enhance rapport should continue throughout the interview. Support the informant through verbal and non-verbal techniques which indicate that their responses are valued. Informants may sometimes recount experiences which upset them or stir other emotions. When an informant is becoming distressed try pausing the interview or changing the topic and possibly returning to the sensitive issue at a later point. If the informant is clearly becoming very distressed you should probably terminate the interview.

There may be a stage in an interview when your informant does not answer a question. If there is a silence or if they shake their head, the informant may be indicating that they have not understood your question, or simply do not know the answer. They might be confused as to the format of the answer expected: is it a 'Yes / No' or something else (Minichiello et al. 1995, p. 93)? In these cases try restating the question, perhaps using an alternative wording or providing an example. You should always be prepared to elaborate on a question. It is important to remember, however, that choosing not to respond is the informant's right. If the informant refuses to answer, and says so, you should not usually press them. They may have chosen not to answer because the question was asked clumsily or insensitively (or for some other reason—if the question dealt with sensitive commercial matters, for instance). If you prepare your questions carefully you should avoid this sort of problem and the consequent loss of empathy and data.

As an interviewer, you should also learn to distinguish between reflective silence and non-answering. Robertson cautions, 'Do not be afraid of silences. Interviewers who consciously delay interrupting a pause often find that a few seconds of reflection leads interviewees to provide the most rewarding parts of an interview ... There is no surer way of inhibiting interviewees than to interrupt, talk too much, argue, or show off your knowledge (Robertson 1994, p. 44).

It is important to allow time for the informant to think, meditate and reflect before they answer a question. It is also important to be patient with slow speakers or people who are not entirely host-language fluent. Resist any temptation to finish peoples' sentences for them. Supplying the word which an informant is struggling to find may seem helpful at the time, but it interrupts them and inserts a term they might not have ordinarily used. In some instances, such as in 'corporate interviews', non-answering may relate to commercial confidentiality or the protection of information which is being kept secret for other reasons. In such instances, a non-answer also becomes data.

Closing the interview

Do not allow rapport to dissipate at the close of an interview. It is critical to maintain rapport—especially if you intend to re-interview the informant. You must prepare for the closure of an interview otherwise the ending can be clumsy. Because an interview establishes a relationship within which certain expectations are created, it is better to indicate a sense of continuation and of feedback and clarification than to end the interview with an air of finality.

Try not to rush the end of an interview. At the same time do not let an interview 'drag on'. There is an array of verbal and non-verbal techniques for closing interviews (Box 4.10). Of course, non-verbal versions should be accompanied by appropriate verbal cues otherwise you could appear quite rude. The most critical issue in closing an interview is to express not only thanks but also satisfaction with the material which was collected. For example you might say: 'Thanks for your time. I've got some really useful/insightful information from this interview'. Not only is gratitude expressed this way, but the informant is made aware that the process has been useful, and that their opinions and experiences have been valued.

Box 4.10: Techniques for closing interviews

Four types of verbal cue:

- Direct announcement
  'Well, I have no more questions just now.'
- Clearing-house questions
  'Is there anything else you would like to add?'
- Summarising the interview

'So, would you agree that the main issues according to you are ....?'
- Making personal inquiries and comments
  'How are the kids?' or *'If you want any advice on how to oppose .... just ring me.'*

Six types of non-verbal cue:
- looking at your watch
- putting the cap on your pen
- stopping or unplugging the tape recorder
- straightening your chair
- closing your notebook
- standing up and offering to shake hands.

***Source*: adapted from Minichiello et al. (1995, pp. 94–8).**

RECORDING AND TRANSCRIBING INTERVIEWS

Interview recording, transcription, and fieldnote assembly are referred to as the mechanical phases of the interview method. These are the steps through which the data are collected, transformed and organised for the final stages of analysis.

Recording

Audio-tape recording and note-taking are the two main techniques for recording an interview. Other less commonly used techniques in geography include video recording, compiling records of the interview after the session has ended and using cognitive maps. Both tape recording and note-taking have associated advantages and disadvantages, as will become clear in the discussion to follow. Therefore, a useful strategy of record keeping is to combine note-taking and audio-tape recording.

The records of an interview should be as close to complete as possible. A tape recorder will help compile the fullest recording (Whyte 1982, pp. 117–8). Interviewers who use note-taking would need excellent shorthand writing skills to produce verbatim records. However, the primary aim in note-taking is to capture the gist of what was said.

Audio or video tape recording can allow for a natural conversational interview style because the interviewer is not preoccupied with taking notes. Instead, you can be a more attentive and critical listener. Tape recording is also preferable to note-taking because it allows you more time to organise the next prompt or question, and to maintain the conversational nature of the interview. The note-taking researcher can be so engrossed in note-taking that they can find themselves unprepared to ask the next question. Note-takers can also miss important movements, expressions and gestures of the informant while they are hunched over and scribbling at a furious pace (Whyte 1982, p. 118). This all undermines rapport and detracts from attentive listening.

On the other hand, a tape recorder may sometimes inhibit an informant's responses because the tape recorder serves as a reminder of the formal situation of the interview (Douglas 1985, p. 83). Informants may feel particularly vulnerable because someone might recognise their voice if the recording was to be aired publicly. Opinions given by the informant on the 'spur of the moment' become fixed indelibly on tape and have the potential to become a permanent public record of the informant's views. This may lead to the informant being less forthcoming than they would have been if note-taking had been used. Some informants become comfortable with a tape recorder as the interview progresses, but others do not. If you find the latter situation to be the case, consider stopping the tape and reverting to note-taking.

If you use a tape recorder, place it somewhere that is not too obvious without compromising the recording quality. The use of long playing tapes will diminish your concern about whether the tape has stopped and minimise the interruption associated with changing tapes. Take care when using a tape recorder not to be lulled into a loss of concentration by the feeling that everything is being recorded safely. There may be a technical failure. You can maintain concentration and avoid the problem associated with tape failure by taking some written notes. If you are taking notes, there is little likelihood of mental wandering. Everything is being listened to, interpreted and parts of it written down, demanding that you maintain concentration. I find this particularly important if I am conducting the second or third interview of a long day's fieldwork.

Because an audio-tape recorder does not keep a record of non-verbal data, non-audible occurrences such as gestures and body language will be lost unless you are also using a video recorder or taking notes. If an informant points to a wall map and says: 'I used to live there', or if they say: 'The river was the colour of that cushion', then the tape recording will be largely meaningless without some written record. These written notes can be woven into the verbal record during the transcription phase (described later in this chapter). Written notes also serve as a back-up record in case of technical failures. Overall, then, a strategic combination of both tape recording and note-taking can provide the most complete record of an interview with the least threat to the interview relationship.

Transcribing the data

Whether using a tape or notes, the record of an interview is usually written up to facilitate analysis. Interviews produce vast data sets which are next to impossible to analyse if they have not been converted to text. A transcript is a written 'reproduction of the formal interview which took place between researcher and informant' (Minichiello et al. 1995, p. 220). The transcript should be the *best possible record* of the interview, including descriptions of gestures and tone as well as the words spoken. The name or initials of each speaker should precede all text in order to identify the interviewer(s) and informant(s). Tape counter numbers at the top and bottom of each page of the transcript enable quick cross-referencing between the **transcript file** and any audio or video tapes of the interview. Converting interview to text is done either through a reconstruction from hand-written notes, a transcription of an audio or video tape, or through the use of voice recognition computer software (See Box 4.11). Issues particularly pertinent to the reconstruction of note-taking-based interviews are outlined below.

Box 4.11: Voice recognition computer packages and interviews

Computing packages have been developed which convert the spoken word into computer text. Packages such as NaturallySpeaking by Dragon Systems which convert text at the rate of 150 or 160 words per minute do not require that each word be enunciated separately. However, these systems will only convert the speech of a single speaker. Each system has to be 'trained' to understand a single 'master's voice'.

The success of these packages for converting interview data has been mixed. The researcher has to simultaneously listen to a tape and verbally repeat the informant's contributions to enable the system to convert the data. Gestures and indications of intonation have to be typed into the document manually. The other major limitations of current voice recognition software include the cost of the packages and the need for very powerful personal computers and a host of computer add-ons such as sound cards. Nonetheless, list-serve reports lodged by researchers have claimed accuracy rates as high as 95% and better systems are on the horizon. It should however be noted that serious investigation of methodological issues which may surround voice recognition software has barely begun.

Interview notes should be converted into a typed format preferably on the same day as the interview. If there were two or more interviewers it is a good idea to compile a combined reconstruction of what was said using each researcher's notebook. This will improve the breadth and depth of coverage. The final typed record will comprise some material recalled verbatim and summaries or approximations of what was said.

Taped interviews should also be transcribed as soon as possible after the interview. Transcription is a very time consuming, and therefore resource intensive, task (Whyte 1982, p. 118). On average, most interviews take four hours of typing per hour of interview. Transcription rates vary according to a host of variables such as typist skill, the type of interview, the informant and the subject matter. You can facilitate transcription by using a purpose-built transcribing recorder. You should transcribe your own interviews for two main reasons. First, since you were present at the interview, you are best placed to reconstruct the interchange. You are aware of non-audible occurrences and therefore know where such events should be inserted into the speech record. You are also better able to understand the meaning of what was said and less likely to misinterpret the spoken words. Second, transcription, although time consuming, does enable you to engage with the data again. Immersion in the data provides a preliminary form of analysis.

While there is no accepted standard for symbols used in transcripts, some of the symbols commonly used are set out in Box 4.12.

Box 4.12: Symbols commonly used in interview transcripts

Symbol	*Meaning*
//	Speaker interrupted by another speaker or event: //phone rings//
:	Also used to indicate an interruption
KMD	The initials of the speaker, usually in CAPS and bold
—	When used at the left margin refers to an unidentified speaker
Ss	Several informants who said the same thing
E	All informants made the same comment simultaneously
...	A self-initiated pause by a speaker
.... or	Longer self-initiated pauses by a speaker
-	Speech which ended abruptly but without interruption
( )	Sections of speech, or a word, which can not be deciphered
(jaunty)	A best guess at what was said
(jaunty/journey)	Two alternative best guesses at what was said
*	Precedes a reconstruction of speech which was not taped
(...)	Material which has been edited out
But <u>I didn't want to</u>	Underlined text indicates stressed discourse
I got nothing	Italicised text indicates louder discourse
[sustained laughter]	Non-verbal actions, gestures, facial expressions
[hesitantly]	Background information on the intonation of discourse

Once completed, the transcript should be given a title page stating the informant's name (or a code if there are concerns of confidentiality), the number of the interview (for example, first or third session), the researcher(s) name(s) (i.e., who carried out the interview), the date of the session, the location, duration of interview, and any important background information on the informant or special circumstances of the interview.

The transcript can be given to the informant for vetting and authorising (see Chapter 3). This will normally improve the quality of your record. This process of **participant checking** continues the involvement of the informants in the research process and provides them with their own record of the interview. You might choose to circle or underline key words or phrases in the transcript. These may be quotes which demonstrate a particular point which could be presented as evidence in a final report on the research.

Assembling fieldnote files

Assembling interview records marks the beginning of the analysis proper. This begins with a critical assessment of the interview content and practice and is followed by formal preparation of interview logs. To my mind the best and most recent explanation of assembling fieldnote files is that by Minichiello et al. (1995, pp. 214–46). In the wide margins of the transcript file you can make written annotations. Comments that relate to the practice of the interview, such as the wording of questions and missed opportunities to prompt, should be placed in the left margin. These annotations and other issues concerned with contact, access, ethics and overall method, should be elaborated upon in a **personal log** (Box 4.13). The right margin of the transcript file can be used for annotations on the substantive issues of the research project. These comments, which generally use the language and jargon of social science, are then elaborated upon in the **analytical log**. The analytical log is an exploration and speculation about what the interview has found in relation to the research question (Box 4.13). It should refer to links between the data gathered in each interview and the established literature or theory.

Box 4.13: Fieldnote files

Transcript file	*Personal log*	*Analytical log*
Includes the record of speech, and the interviewer's observations of non-audible data and intonation. Also includes written annotations in the margins on the practice and content of the interview.	Reflection on the practice of the interview. Includes comments on the questions asked and their wording, the appropriateness of the informant, recruitment and access, ethical concerns, and the method generally.	Exploration of the content of the interview. A critical outline of the substantive matters which have arisen. Identification of themes. Reference to the literature and theory. In-progress commentary on the research aims and findings.

Source: adapted from Minichiello et al. (1995, pp. 214–46).

ANALYSING INTERVIEW DATA

Researchers analyse interview data to seek meaning from the data. We construct themes, relations between variables and patterns in the data through content analysis (see Chapter 7). *Content analysis* can be based on a search of either manifest or latent content (Babbie 1992, pp. 318–19). **Manifest content analysis** assesses the visible, surface content of a document such as an interview transcript. An example would be a tally of the number of times the words 'cute' and 'cuddly' are used to describe koalas in interviews with members of the public. This might contribute to a broader assessment of the political significance of culling in areas of koala overpopulation (for example, Muller 1999). Searching interview data for manifest content often involves tallying the appearance of a word or phrase. Computer programs such as ***NUD*IST*** are particularly effective at undertaking these sorts of manifest searches (see Chapter 8).

Latent content analysis involves searching the document for themes. For example you might keep a tally of each instance in which a female has been portrayed in a passive or active role. Latent content analysis of interview texts requires a determination of the underlying meanings of what was said. This determination of meanings within the text is a form of coding.

A *coding system* is used to sort and then retrieve data. For example the text in transcripts of interviews with 'urban managers' could be coded according to the following categories: managerialist perspectives, entrepreneurial perspectives, impact of globalisation, impact of government legislation, democratic concerns, and economic efficiency. Once the sections of all the interviews have been coded, it is then possible to retrieve all similarly coded sections. These sections of text can be amalgamated and re-read as a single file (Box 4.14). This might allow a researcher to grasp the varying opinions on a certain issue and to begin to unravel the general feeling about an issue.

Box 4.14: Coding interview data: five suggested steps

Coding step	*Specific operations: computing / manual versions*
Develop preliminary coding system	Prepare a list of emergent themes in the research. Draw on the literature, your past findings, as well as your memos and log comments. Amend throughout.
Prepare the transcript for analysis	Meet the formatting requirements for the computing package being used. / *Print out a fresh copy of the transcript for manual coding.*
Ascribe codes to text	Allocate coding annotations using the 'Code Text' function of computing packages. / *Place hand-written annotations on transcript.*
Retrieve similarly coded text	Use the 'Retrieve Text' function of computing packages to produce reports on themes. / *Extract and amalgamate sections of text which are similarly coded.*
Review the data by themes	Assess the diversity of opinion under each theme. Cross-referencing themes allows you to review instances where two themes are discussed together. Begin to speculate on relations between themes.

Not every section of text needs to be coded. An interview will include material which is not relevant to the research question, particularly warm-up and closing sections, and other speech focused on improving rapport rather than gathering data. Sections of text can also be multiple-coded. For example in one sentence an informant may list a number of causes of fish kills including open-cut mine run-off, super phosphates, acid sulphate soils and town sewage. This may require that the sentence is attributed four different coding values.

PRESENTING INTERVIEW DATA

Material collected from interviews is rarely presented in its entirety. Most interview data must be edited and (re)presented selectively in research publications. While it is difficult to locate a 'genuinely representative' statement (see Connell 1991, pp. 144–45; Minichiello et al. 1995, pp. 114–15), it is usually possible to indicate the general sense and range of opinion and experience expressed in interviews. One way to indicate this is to present summary statistics of what was said. Computing packages such as *NUD*IST* can help you calculate the frequency with which a particular term or phrase appeared in a document or section of text (see Chapter 8). However, the more common method is through a literal description of the themes which emerged in the interviews (for example, Boxes 4.6, 9.1 and 9.2, and the discussion in Chapter 5 on presenting results).

When describing interview data you must cite transcript files appropriately. For example, in her interview-based honours research on the changing identity of the industrial city of Wollongong, Pearson (1996, p. 62) noted that 'Several respondents asserted that elements excluded by the new identity were of little significance to the overall vernacular identity of Wollongong' (Int.#1, Int.#6 and Int.#7). The transcript citations provided here indicate which of the informants expressed a particular type of opinion. In research publications the transcript citations can indicate the informant's name, number, code, or tape number. Whenever a direct quotation from an informant is presented then a transcript page reference or tape counter number should be provided.

Transcript material should be treated as data. A quotation, for example, ought to be treated in much the same way as a table of statistics. That is, it should be introduced and then interpreted by the author. The introduction to a quotation should offer, if it has not already been provided, some background on the informant. It is important that readers have some idea of where an informant is 'coming from'; information about their role, occupation or status is important in this regard. Also important, as Baxter and Eyles (1997, p. 508) point out, is 'some discussion of why particular voices are heard and others are silenced through the selection of quotes'. (See Chapter 3 for a discussion of this sort of transparency in research reporting.) Quotations should be discussed in relation to, and contrasted with, the experiences or opinions of other informants. Statements of opinion by an informant should also be assessed for internal contradiction. Finally, a quote cannot replace a researcher's own words and interpretation. As the author you must explain clearly what theme or issue a quotation demonstrates.

Knowledge is a form of power. The accumulation and ownership of knowledge is an accumulation of power: power to effect change, power to support arguments or to construct proofs. In most interviews information and knowledge flow from the informant to the researcher. The researcher accumulates this knowledge and ultimately controls it. There is a host of strategies and guidelines to which researchers can adhere to reduce the potential political and ethical inequities of this relationship (see Chapters 2, 3 and 9). In terms of data presentation it will sometimes be important that an informant's identity be concealed. Pseudonyms or interviewee numbers have been used by geographers to disguise the identity of their informants where it has been thought that disclosure could be harmful. Naming an informant, and particularly their direct association to a quotation, could be personally, professionally or politically harmful. Researchers must be very careful when they deploy data they have collected. Interviewers are privileged with insights into people's lives. Research deals and promises should be respected. In this way the integrity of the researcher, and of the entire research community, will be enhanced (Hay 1998).

Finally, the presentation of interview-based research must contain an accessible and transparent account of how the data were collected and analysed (Baxter and Eyles 1997, p. 518). This account should outline the subjectivity of the researcher, including their biases or 'positioned subjectivity' (see Chapters 2, 3 and 9). Some indication should also be given on what procedures were used for selecting interview excerpts for presentation and of how instances of shared or divergent opinion were determined by the researcher. As we have already seen from the discussion in Chapter 3, it is only through transparent accounts of how interview-based research was undertaken that the trustworthiness and wider applicability of the findings can be assessed by other researchers.

CONCLUSION

The rigour of interview-based research is enhanced through adequate preparation, diverse input and verification of interpretation. Being well informed and prepared will allow a deeper understanding of the 'culture' and discourse of the group(s) you study. You can then formulate good questions and enhance levels of rapport between you and your informants. You should also purposely seek out diversity of opinion. Interviewing more than one informant from each study group will begin to draw out and invite controversy or tensions. An opinion from one informant should never be accepted as demonstrative of group opinion unless it is shown to be the case. Finally, some means of verifying your interpretations of interview data are necessary (for example participant checking, peer checking and cross reference to documentary material).

Interviews bring people 'into' the research process. They provide data on peoples' behaviour and experiences. They capture informants' views of life. Informants use their own words or vernacular to describe their own experiences and perceptions. Kearns (1991, p. 2) makes the point that 'there is no better introduction to a population than the people themselves'. This is what I find to be the most refreshing aspect of interview material. Transcribed interviews are wholly unlike other forms of data. The informant's non-academic text reminds the researcher and the reader of the research of the lived experience which has been divulged. It reminds geographers that there are real people behind the data.

KEY TERMS

aide mémoire
analytical log
critical inner dialogue
funnelling
informant
interview guide
interview schedule
latent content analysis
life history
manifest content analysis
*NUD*IST*
oral history
participant checking
personal log
primary question
prompt
pyramid structure
rapport
semi-structured interview
structured interview
transcript file
unstructured interview

REVIEW QUESTIONS

1 Select one of the three questions below and spend about 15 minutes constructing an interview schedule for a hypothetical five-minute interview with one of your colleagues. Use a mix of primary question types and prompts. Think about the overall structure of your schedule and provide a sense of order to the way the issues are covered. Try to imagine how you will cope if the interviewee is aggressive, very talkative or non-communicative. Will your schedule still work?

- a Most of us would agree that a greater use of public transport is an environmentally and economically sound goal. However most of us would personally prefer to use a private car and only pay lip service to such noble goals. Why?
- b Beach activity is decidedly spatial. Performances are expressive and behaviour is at times territorial.
- c The re-integration of the differently abled into 'normal society' is a noble ideal. However, this integration will always be confounded by the organisation of public space and the reactions of the able-bodied when the differently abled are in public space.

2 Conduct two semi-structured in-depth interviews with someone of an older generation than yourself. It could be an older relative (however, do not interview a sibling or parent). Limit both interviews to approximately thirty minutes. Construct an interview guide which operationalises key concepts in the following research question: 'Ours is a patriarchal society. We are often told, however, that the society of our parents and grandparents was structured by an even more restrictive and oppressive system of sexism. Investigate how the opportunities, resources and experiences differed according to gender for earlier generations. Pay particular attention to gender variations in the use of, and access to, space.'

3 Differentiate between an oral history and a life history.

4 Devise a list of rapport strategies you could use if you were to interview an older relative not well known to you. Consider the preliminary, contact, warm-up and closing phases of the interview.

SUGGESTED READING

Baxter, J. and Eyles, J. 1997, 'Evaluating qualitative research in social geography: establishing "rigour" in interview analysis', *Transactions of the Institute of British Geographers*, vol. 22, no. 4, pp. 505–25.

Douglas, J.D. 1985, *Creative Interviewing*, Sage, Beverly Hills.

Edwards, J.A. and Lampert, M.D. (eds) 1993, *Talking Data: Transcription and Coding in Discourse Research*, Lawrence Erlbaum Associates, Hillsdale, New Jersey.

Findlay, A.M. and Li, F.L.N. 1997, 'An auto-biographical approach to understanding migration: the case of Hong Kong emigrants', *Area*, vol. 29, no. 1, pp. 34–44.

Kearns, R. 1991, 'Talking and listening: avenues to geographical understanding', *New Zealand Journal of Geography*, vol. 92, pp. 2–3.

Minichiello, V., Aroni, R., Timewell, E. and Alexander, L. 1995, *In-Depth Interviewing: Principles, Techniques, Analysis*, 2nd edn, Longman Cheshire, Melbourne.

Oakley, A. 1981, *From Here to Maternity: Becoming a Mother*, Penguin, Middlesex.

Robertson, B.M. 1994, *Oral History Handbook,* Oral History Association of Australia, Adelaide.

Schoenberger, E. 1991, 'The corporate interview as a research method in economic geography', *Professional Geographer*, vol. 43, no. 2, pp. 180–89.

Tremblay, M.-A. 1982, 'The key informant technique: a non-ethnographic application', in *Field Research: A Sourcebook and Field Manual*, ed. R.G. Burgess, Allen and Unwin, London.

Whyte, W.F. 1982, 'Interviewing in field research', in *Field Research: A Sourcebook and Field Manual*, ed. R.G. Burgess, Allen and Unwin, London.

5

Focussing on the Focus Group

Jenny Cameron

Contents

Chapter overview

An investigation of community responses to literature on environmental sustainability (Myers and Macnaghten 1998), a study of rapid social and economic change in non-metropolitan regions (Gibson, Cameron and Veno 1999), an examination of the construction of identity through shopping (Jackson and Holbrook 1995), an exploration of the daily lives of young single-parent African–American women (Jarrett 1994)—all of these are examples of research projects which employ focus groups as a means of disentangling the complex web of relations and processes, meaning and representation, that comprise the social world. With the shift to more nuanced explorations of people–place relationships in geography the focus group method has been recognised increasingly as a valuable research tool.

Focus groups can be exhilarating and exciting, with people responding to the ideas and viewpoints expressed by others, and introducing you, the researcher, and other group members to new ways of thinking about an issue or topic. This chapter discusses the diverse research potential of focus groups in geography, outlines the key issues to consider when planning and conducting successful focus groups, and provides an overview of strategies for analysing and presenting the results.

WHAT ARE FOCUS GROUPS?

The focus group method involves a small group of people discussing a topic or issues defined by a researcher. Briefly, a group of between six and ten people sit facing each other around a table, the researcher introduces the topic for discussion and then invites and moderates discussion from group members. A session usually lasts for between one and two hours (you might see parallels here with university tutorial group meetings!).

Interaction between members of the group is a key characteristic of this research method, and it is that which helps differentiate it from the interview method, where interaction is between interviewer and interviewee. The group setting is generally characterised by dynamism and energy as people respond to the contributions of others. One comment, for example, can trigger a chain of responses. This type of interaction has been described as the 'synergistic' effect of focus groups and some propose that it results in far more information being generated than in other research methods (Berg 1989; Stewart and Shamdasani 1990). In the focus group excerpt in Box 5.1, for example, the discussion shifts from family farming practices, to people's commitment to an area, to ways of working with government, to projects that address environmental degradation. Yet as the farmer points out at the conclusion of this excerpt, the speakers all highlight the effect that taking a long-term approach has on economic, environmental and community practices.

Box 5.1: The 'synergistic' effect of focus groups

Farmer A:	Where we make a mistake in business is in thinking of tomorrow. The family approach is what's happening to the next generation. It's a much longer term approach. I'm more interested in investing my resources for the next generation and therefore you build a solid business.
Farmer B:	My attitude is that I'm the tenant in time.
Farmer A:	The custodian.
Farmer B:	Yeah, the custodian. My father gave it to me and I'll hand it on to the next generation. And people say you could sell it and make lots of money but that just doesn't come into the equation. The thought of selling it and leaving the good life—the kids probably will. And I think there are an awful lot of farmers with that attitude. And I think it has probably in lots of ways been to our detriment. We could use that asset and make more money—as if money is the most important thing.
Consultant:	I think that's right. I think one of the important reasons there have been successes and perhaps less problems here is that even though we have all identified lots of problems, we are really committed to this community and making it better. And I think there are an amazing range of people that do choose to live here—they don't have to—but choose to live here and [have] invested huge amounts of time and energy. And I also think this community, just thinking back to my experience, that it's really open to working with whatever government is in at the time and turning the rules or the policies or the dollars that are around for the best here. Like local government saying we don't want yet another regional development board but we will have the money and this is our structure and this is what we'll do. I think there has been some creative use of government money and good partnerships and also just that huge commitment, that energy to make it work.
Manager:	You mention our successes and I think one of the unheralded things we've done really well is look towards the sustainability of the whole area from land management which underpins our whole economy. Because we've poured irrigation water onto this country for years and years and we've never really looked at the repercussions: the drainage problems, the salinity problems. And I think in recent years, in the last fifteen, twenty years, that's really been addressed—the work that's gone into it by some very dedicated people and I think that message has gone across to virtually all land holders in the area. With the advent of some major arterial drains, community drains, the cooperation—the cooperatives virtually that have been formed to bring this into being, really will underpin the future of our economy and the management of our natural resource, which is absolutely vital to the future of our farmers and businesses etc.
Consultant:	And a lot of that work's been voluntary.
Farmer A:	It all comes back to the notion that it's the next generation. It's a different approach.

Source: videotape excerpt from focus group conducted by Katherine Gibson, Jenny Cameron and Arthur Veno, Shepparton, Victoria, 5 June 1997 (see Gibson, Cameron and Veno 1999).

The interactive aspect of focus groups also provides an opportunity for people to explore different points of view, and formulate and reconsider their own ideas and understandings. Kitzinger (1994, p. 113) describes this form of interaction in the following terms: '[p]articipants do not just agree with each other. They also misunderstand one another, question one another, try to persuade each other of the justice of their own point of view and sometimes they vehemently disagree'. For researchers who are interested in the socially constructed nature of knowledge this aspect of focus groups makes them an ideal research method; the multiple meanings that people attribute to places, relationships, processes and events are expressed and negotiated, thereby providing important insights into the practice of knowledge production.

A second characteristic is the pivotal role of the researcher, who promotes group interaction and focuses the discussion on the topic or issue. The researcher draws out the range of views and understandings within the group, and manages—and sometimes even encourages—disagreement between participants (Myers 1998). By comparison, in an observation situation, the researcher may have a more 'hands off' role (see Chapter 6).

Initially focus groups can be extremely challenging for researchers who are new to the process. They are, however, well worth it. In focus groups the diversity of processes and practices that make up the social world and the richness of the relationships between people and places can be addressed and explored explicitly. A not inconsequential consideration is that group members almost invariably enjoy interacting with each other, offering their points of view and learning from each other. Researchers also find the process refreshing (for example, see the discussion by two skeptical anthropologists in Agar and MacDonald [1995]).

USING FOCUS GROUPS IN GEOGRAPHY

Focus group discussions—or focused interviews, as they were originally known (Merton 1987)—were used by sociologists in the United States during World War II to examine the impact of wartime propaganda and the effectiveness of military training materials (Merton 1987; Morgan 1997). Although this work resulted in several sociological publications on the technique, focus groups were neglected by social scientists in the post-World War II period in favour of one-to-one interviews and participant observation (Johnson 1996). It was in the field of market research that the focus group method found a home. Since the 1980s there has been renewed interest in the technique among social scientists and this has led to considerable diversity in the practice of focus group research (Lunt and Livingstone 1996; Morgan 1997). During the war years, for example, focus groups were specifically used in association with quantitative techniques to ensure that results could be generalised (Merton 1987). While this practice continues, some researchers are now more interested in the interactive dynamic which results in an ebb and flow of ideas and understandings. Where once focus groups were primarily a data-gathering tool today they are increasingly used in 'more critical, politicized, and more theoretically driven research contexts' (Lunt and Livingstone 1996, p. 80), exploring, for instance, the **discourses** which shape practices of everyday life, the ways in which meanings are reworked and subverted, and the creation of new knowledges out of seemingly familiar understandings. The range of uses and purposes of focus groups is evident in geographic research employing the technique.

Geographers have used focus groups to collect information. Zeigler, Brunn, and Johnson (1996) used them to find out about peoples' responses to emergency procedures during a major hurricane. They claim that the focus group technique provided insights that might not have been revealed through methods like questionnaires or individual interviews. As a consequence they were able to recommend important refinements to disaster plans. Burgess (1996) has also used focus groups, in combination with participant observation, to obtain information about factors that inhibit visits to, and use of, woodlands. Her findings have contributed to the development of landscape design and management strategies to enhance the use of woodlands.

One concern of some researchers involved in data gathering is that because of the relatively small numbers of participants in focus groups the findings are not applicable to a wider population (for a discussion of this sort of issue, see Chapter 3). Combining focus groups with quantitative techniques is an extremely useful way of dealing with this issue. A survey questionnaire, for instance, might be administered to a random sample of the population from which the focus group was drawn to test the generalisability of the insights gained from the group discussions. Quantitative methods can supplement focus groups in other ways. Preliminary surveys are sometimes helpful in identifying focus group members or the topics for detailed focus group discussion.

Conversely, focus groups can supplement quantitative research. They have been used to generate questions and theories to be tested in surveys (Goss and Leinbach 1996), to refine the design of survey questionnaires (Jackson and Holbrook 1995), and to follow up the interpretation of survey findings (Goss and Leinbach 1996), particularly where there seem to be contradictory results (Morgan 1996). It is, however, entirely appropriate to use focus groups as the sole research method rather than in combination with other research techniques.

For geographers interested in the process of knowledge production focus groups are an excellent research tool. Robyn Longhurst (1996) is a geographer from New Zealand/Aotearoa interested in the absence of a language to talk about pregnancy. She has employed focus groups as a forum in which pregnant women could converse and interact. The narratives, accounts, anecdotes and explanations offered by these women provided Longhurst (1996) with insights into a new discursive landscape of pregnancy. Similarly, Gibson, Cameron, and Veno (1999) have been concerned to not just reproduce a knowledge of the problems and difficulties confronting rural and non-metropolitan communities in Australia, but to reshape understandings so that new responses might be engendered. The seemingly isolated instances of innovation that several focus group members could readily recall provoked other participants to think of additional examples. The beginnings of a body of knowledge on regional initiative began to emerge through these discussions.

In a report of their Indonesian research on individual and household strategies related to the allocation of land, labour, and capital, Goss and Leinbach (1996) also highlight the collective rather than individual nature of knowledge production. By interacting with other focus group members Javanese villagers developed new understandings of their social conditions. Indeed, Goss and Leinbach argue that 'the main advantage of focus group discussions is that both the researcher *and* the research subjects may simultaneously obtain insights and understanding of particular social situations *during* the process of research' (Goss and Leinbach 1996, pp. 116–17, emphasis in original). For geographers who are committed to the idea that research can be used to effect social change and empower 'the researched', the potential for focus groups to create and transform knowledges and understandings of researchers *and* participants is compelling (see also Johnson 1996; Swenson, Griswold and Kleiber 1992).

The focus group method has an important contribution to make to geographic research. It is a highly effective vehicle for exploring the nuances and complexities associated with people-place relationships. The material generated in focus groups can add important insights to work that seeks to describe and document the social world. But focus groups serve not just to 'mine', 'uncover', and 'extract' existing knowledges (Gibson-Graham 1994); they can also contribute to the development and construction of new knowledges and understandings for both researcher and 'researched'.

PLANNING AND CONDUCTING FOCUS GROUPS

Given that the focus group method can be used for a range of research purposes in geography, there will be some variation in how groups are organised and conducted. There are, however, basic principles and methodological and theoretical issues that need to be considered. To be sure, the success of a focus group depends largely on the care taken in the initial planning stage.

Selecting participants

Selecting participants for focus groups is critically important. Generally, participants are chosen on the basis of their experience related to the research topic. Swenson, Griswold and Kleiber (1992, p. 462) refer to this as '**purposeful sampling**', as opposed to the random sampling that characterises many quantitative studies (see Chapter 3 for a discussion of participant selection). Burgess' (1996) study is a good example of purposeful sampling (sometimes known as purposive sampling). In work intended to ascertain the perceptions among different social and cultural groups of crime and risk in woodlands she selected women and men of varying age, stage in the life cycle, and ethnicity to participate in focus groups. In another study (Casey et al. 1996) of local perspectives on potential strategies to address agricultural pollution in the Minnesota River Basin, people from the area involved in different aspects of farming were invited to participate. Groups were made up of farmers—who varied in age and gender, and size and type of farm—and local staff from agriculturally-based government agencies and non-profit groups.

Composition of focus groups

Should people with similar characteristics participate in the same group or should groups comprise members with different characteristics? This decision will be largely determined by the purpose of your research project.

Holbrook and Jackson (1996), for example, sought to address issues of identity, community and locality and thought it appropriate to group together people with characteristics like age and ethnicity in common (see also Burgess 1996). Other researchers have noted that discussion of sensitive or controversial topics can be enhanced when groups comprise participants who share key characteristics (Hoppe et al. 1995; O'Brien 1993). Goss and Leinbach (1996) were interested in the social relations involved in family decision-making and deliberately chose to conduct mixed gender groups. The different knowledges, experiences, and perspectives expressed by women and men became an important point of discussion.

Another consideration is whether people already known to each other should participate in the same group. In some research, particularly place-based research, it may be unavoidable that group members are acquainted. Here confidentiality can be an issue as participants tend to over-disclose information about themselves in focus groups, rather than under-disclose (Morgan and Krueger 1993). One strategy for dealing with over-**disclosure** is to outline fictional examples and ask group members to speculate and comment on these. In groups they ran in Indonesia, Goss and Leinbach (1996) provided details of three fictional families and asked group members to discuss which of the families would be most likely to accumulate capital. This shifted the emphasis away from the specifics of the lives of group members, while still enabling discussion of family strategies.

Participants can also be asked to treat discussions as confidential. As this cannot be guaranteed, it is appropriate to remind people to disclose only those things they would feel comfortable about being repeated outside the group. Of course, you should always weigh up whether a topic is too controversial or sensitive for discussion in a focus group and is better handled through another technique, like individual in-depth interviews. (Most universities now have ethics committees to ensure that researchers carefully manage material from focus groups and other qualitative research methods. For more on this see Chapter 2.)

Size and number of groups

The size of each group and the number of groups are other factors to be considered in the planning stage. Too few participants per group—fewer than four—limits the discussion, while too many—more than ten—restricts the time for participants to contribute.

While one rule of thumb is to hold three to five groups, this will be mediated by a number of factors (Morgan 1997) such as the purposes and scale of the research and the heterogeneity of the participants. A diverse range of participants is likely to necessitate a larger number of groups. Burgess (1996), for instance, conducted thirteen focus groups with people of varying age, stage in the life cycle and ethnicity while Longhurst (1996) held five groups with women living in Hamilton, New Zealand/Aotearoa who were pregnant for the first time.

The structure of the focus group is also a factor to consider when planning your research. When less standardised questions are used and when there is a relatively low level of researcher intervention and moderation more groups are needed, as both these factors tend to produce greater variability between groups (Morgan 1997). Time, cost and availability of participants may also limit the number of groups that can be held. Finally, the overall research plan, especially whether focus groups are the sole research tool or one of a number of tools, will also affect decisions about the number of groups convened.

Recruiting participants

The strategy you use to recruit participants for focus groups will depend largely on the type of participants you select for inclusion. Gibson, Cameron and Veno (1999) recruited business and community leaders in two regions by reading reports of community events in local newspapers and targeting managers of key government and non-government agencies. After an initial set of potential participants were contacted, a snowball **recruitment** technique was used to contact other people. In this technique existing group members identify additional people to be included in the study (the snowball recruitment technique is also discussed in Chapter 3). A preliminary phone conversation quickly established whether nominees were interested and able to attend. This was followed by a letter with more information about the project. A few days before the focus groups were held, participants were telephoned again to re-confirm their participation. Twelve people were invited to attend each group to allow for cancellations due to illness, last-minute change of plans and so on (several people from each group did cancel).

After an unsuccessful attempt to recruit participants for their study by advertising in local newspapers and writing letters to local organisations, Holbrook and Jackson (1996) went directly to the places where potential participants were likely to meet and socialise, such as community centres, homes for the elderly, play groups and clubs. Managers or convenors of the centres helped set up the groups, or the researchers visited venues and invited people to participate. Once people had been involved in a focus group, news of the project spread by word of mouth and other people were recruited easily. Like Holbrook and Jackson, researchers need to think strategically about how best to locate potential participants (see also Burgess, Limb and Harrison 1988a).

Questions and topics

Before conducting focus groups, give thought to the questions or topics for discussion. This involves not only the general content of questions or selection of issues for discussion, but also the wording of questions and issues, identification of key phrases that might be useful, the sequencing and grouping of questions (see Chapter 4 for additional material on question order), strategies for introducing issues, and the links that might be important to make between different questions or issues.

One way to proceed is to devise a list of questions. Swenson, Griswold, and Kleiber (1992) developed a list of twenty questions to act as **probes** for focus groups comprising rural journalists. Another list of twenty questions was used in separate focus groups with community development leaders. Instead of a list of questions, Holbrook and Jackson (1996) identified six themes related to the experience of shopping and then used these to develop specific questions that were raised spontaneously and that fitted with the flow of the discussion. Burgess (1996) preceded each focus group with a walk through a woodland and then introduced for discussion five primary themes related to elements of the walk. Before a series of focus groups they conducted, Gibson, Cameron and Veno (1999) asked each participant to prepare a brief two-minute summary of the major social and economic changes they thought had occurred in their region over the last twenty years. The similarities and differences revealed through these statements provided the basis for discussion.

You should take care about letting people know in advance what the questions or topics will be. If attendance or discussion is likely to be enhanced by providing this information then it may be appropriate. Sometimes, however, it might be necessary for you to paint a very broad picture. For example it might be more judicious to let a group of men know that you are interested in how they manage the interrelation between work, recreation and home than to tell them you are interested in contemporary negotiations of masculinity (provided of course that you do want to know about masculinity in work, recreation and home environments). (See Chapter 2 for a consideration of the ethical dimensions of this sort of approach.) Another lesson that can be drawn from this paragraph is to be sure to use language that participants will understand when you are providing them with advice on the themes of your research.

Generally, questions or topics should allow for discussion of between one and two hours. With very talkative groups it might be necessary to intervene and move the discussion on to new topics. Alternatively, if a hierarchy of questions or themes has been established in the planning stage then it may be appropriate to allow the group to focus on the more important areas of discussion. With less talkative groups you may need to introduce additional or rephrased questions and prompts to help draw information out and open up the discussion. These should be thought about in the preparation stage.

Another issue to consider is whether questions and topics will be standardised across all focus groups involved in your study or whether new insights from one group will be introduced into the discussions of the next. In many qualitative research situations it may be appropriate to incorporate material from earlier groups, but this should be determined by referring to the project aim. Information that might identify people who attended earlier groups must not be revealed to subsequent groups.

As well as running meetings with several groups, you may find it useful and appropriate to have each group meet more than once. Burgess, Limb and Harrison (1988a and b) ran in-depth discussion groups that met each week for six weeks to explore individually and collectively held environmental values. Although this group method is slightly different from the focus group method—it draws on the psychotherapeutic tradition and places an emphasis on the exploration of feelings and experiences—it does not preclude focus groups from meeting more than once.

Multiple focus groups may be a particularly useful strategy when participants are being asked to explore new and unfamiliar topics or to think about an apparently familiar topic in a new way (such as Longhurst's [1996] research on a new language of pregnancy). They may also be appropriate as a way of developing trust between the researcher and research participants. For instance, when researching the experiences of single mothers I met several times with one group of teenagers who were very wary of talking with people associated with educational, medical and media institutions (Cameron 1992).

Conducting focus groups

Generally, focus groups are best held in an informal setting that is easily accessible to all participants. The rooms of local community centres, libraries, churches, schools and so on are usually ideal. The setting should also be relatively neutral: for example it would not be advisable to convene a focus group about the quality of service provided by an agency in that agency's offices. Food and drink can be offered to participants when they arrive to help them relax, but alcohol should never be provided. It is also helpful to give out name tags as participants arrive.

There has been much written about the ideal focus group **facilitator** or **moderator** (for example, Morgan 1997; Stewart and Shamdasani 1990). In academic research it is often the researcher, who is familiar with the aim of the research project and the purpose of the focus groups, who is best positioned to fill this role. To gain some confidence and familiarity with the process, a less experienced researcher might initially take the role of note-taker while a more experienced researcher facilitates the first groups. Focus groups can also be run with more than one facilitator, and a less experienced researcher might invite a more experienced researcher to take the lead facilitating role.

When a note-taker is present they should sit discretely to one side of the group. The notes, particularly a list of who speaks in what order and a brief description of what they talk about, can be helpful when transcribing any audio-tapes that might be made of the focus group discussion. As the facilitator has to attend to what group members are saying and monitor the mood of the group, they should not take extensive notes, though they may want to jot down a point to come back to in discussion.

It is highly advisable to audio-tape focus groups. The group will usually cover so much material that it is impossible to recall everything that was discussed. In addition, because written material from focus group research generally includes direct quotes to illustrate key points, a transcribable tape recording of the meeting can be very helpful. The quality of the taperecorder and microphone is crucial. Most recorders come with a built-in microphone, but several flat 'desk' microphones placed around the table will ensure that quieter voices are heard on tape and much better sound quality. The audiovisual departments of universities are sometimes excellent places to get advice. Ensure you test the equipment long before the focus group, and also later in the room before group members arrive. Spare batteries and tapes are essential equipment for focus group researchers. Take care to ensure that the setting for the group meeting is quiet enough for discussion to be picked up clearly on tape.

The facilitator usually initiates discussion by giving an overview of the research project and the role of the focus group in the project. The themes or questions for discussion can then be introduced. As group members may be unfamiliar with the focus group technique, a brief summary of how focus groups operate should also be given. Box 5.2 provides an example of a focus group introduction.

Box 5.2: A sample introduction to a focus group session

In this focus group, three researchers acted as facilitators. As people arrived they were greeted by one of the research team, introduced to the other researchers and group members and offered tea or coffee. When all participants had arrived the group was invited to sit around a table. The primary facilitator for this focus group session explained the consent form that was already laid out in front of each person. Participants were asked to read and sign it. The consent forms were passed to one of the researchers, and the session was ready to begin. The researcher acting as the primary facilitator introduced the project:

> Well, I'd like to thank you all for making the time and coming along today and contributing and sharing your knowledge. In this particular project that we're working on, we're looking at how communities negotiate change and the reason that we particularly wanted to look at this community was that it seemed there was a lot of change going on and the community, ummm, seemed to be, ummm—we were interested in how you saw your community handling that change and specifically what we're trying to get is to—in the long term is to generate a set of suggestions for other communities on how to manage and negotiate change. So we're hoping to learn from both the mistakes and the right things you've done. So what we're looking for today is a frank and open discussion about how you see change occurring in your community over the last twenty years and how that has been handled. And a little later on in the session we'll get you to—as we go on through the session we'll ask you specific things to help flesh out answers and issues that might be raised. And anything you feel like contributing or adding to just jump in and have that because these focus groups are to get at what the ideas and issues are as you see them. So I think we mentioned in the initial contact with you that we'd like to start out with a two-minute presentation from each of you as to how you see the critical features of change in your community. So we might start around this way. And if you would introduce yourself and your affiliation as you start.

Once all the group members had made their presentations the primary facilitator opened the discussion up:

> Great, thanks very much. Well that's been really informative to get all those different perspectives. What we'd like to do now is to explore some of these issues. But from here on in the process should change and you should feel free enough to ask, agree, disagree, jump in, put your opinion forward, and so on. And if things get a bit noisy then we'll just jump in and try and get some semblance of order. One of the common themes that runs through what you've all said is that the community fabric has been affected in a really negative way by all the changes that have occurred. And what I'm trying to get at is what could have been done to improve that. So what do you think?

From this point on different group members responded to questions from the researchers, asked each other questions, agreed and disagreed with each other. The topics for discussion flowed as people each contributed adding a slightly different perspective and introducing new ideas. The researchers also asked questions and points of clarification and introduced new areas of discussion.

***Source*: videotape excerpt from focus group conducted by Katherine Gibson, Jenny Cameron and Arthur Veno, La Trobe Valley, Victoria, 19 June 1997 (see Gibson, Cameron and Veno 1999). An example of an introduction is also provided by Myers (1998, p. 90).**

The facilitator moderates discussion by encouraging exploration of a topic, introducing new topics, keeping the discussion on track, encouraging agreement and disagreement, curbing talkative group members and encouraging quiet participants. Examples of the sorts of phrases used by facilitators are outlined in Box 5.3.

Box 5.3: Examples of phrases used in focus group facilitation

- Encouraging exploration of an idea:
  'Does anyone have anything they'd like to add to that?'
  'How do you think that relates to what was said earlier about . . . ?'
  'Can we talk about this idea a bit further?'
- Moving onto a different topic:
  'This is probably a good point to move on to talk about . . . .'
  'Just following on from that, I'd like bring up something we've not talked about yet.'
  'This is an important point because it really picks up on another issue.'
- Keeping on track:
  'There was an important point made over here a moment ago, can we just come back to that.'
- Inviting agreement:
  'Has anyone else had a similar experience?'
  'Does anyone else share that view?'
- Inviting disagreement:
  'Does anyone have a different reaction?'
  'We've been hearing about one point of view but I think there might be other ways of looking at this. Would anyone like to comment on other sorts of views that they think other people might have?'
  'There seems to be some differences in what's been said and I think it is really important to get a sense why we have such different views.'
- Curbing a talkative person:
  'There's a few people who've got something to add at this point, we'll just move onto them.'

'We need to move onto the next topic, we'll come back to that idea if we have time.'

- Encouraging a very quiet person:
  'Do you have anything you'd like to add at this point?'

Source: drawn from discussions in Carey (1994) and Myers (1998) and from personal experience.

Some aspects of facilitation require special comment. The expression and exploration of different points of view is important in focus groups, yet research has shown that groups exhibit a preference for agreement (Myers 1998). The facilitator plays a central role in creating the context for disagreement. This can be done by stating in the introduction that disagreement is normal and expected, by asking directly for different points of view, and by making explicit implied disagreement and introducing it as a topic for discussion (Myers 1998, p. 97). Of course, as facilitator, you should never state that someone is wrong, nor display a preference for one position or another. In the unlikely event that the discussion becomes heated then you should intervene immediately, suggest that there may be no right answer, and move the group on to the next topic.

Very talkative or very quiet participants can be a problem. Talkative people need to be gently curbed, while quiet ones need to be encouraged to participate. Along with the sorts of phrases listed in Box 5.3, your non-verbal signals can be useful. Pointing to someone who is waiting to speak indicates to the talkative person that there are others who need to have a turn. Making frequent eye-contact with the quieter person and offering signs of encouragement, like nodding and smiling when they do speak, is important.

At the conclusion of a focus group you might review key points of the discussion, providing a sense of completion and allowing participants to clarify and correct the facilitator's summary. Group members should always be thanked for taking the time to attend and for their contributions. You should consider doing this in the form of a personal letter to each participant.

ANALYSING AND PRESENTING RESULTS

Since there is always a richness of material, analysing focus group discussions can be as time-consuming as it is interesting. The first step involves transcribing the audio-tapes. A complete **transcript** of the entire discussion can be time-consuming, as one hour of tape usually takes over four hours to transcribe. When a detailed analysis and comparison of groups is to be undertaken a full transcription may be necessary. Generally a partial or abridged transcription (which involves transcribing only key sections of the discussion) will suffice. This is best done as soon as possible after the focus group with the facilitator/s and note-taker working in collaboration to decide which sections should be transcribed. A record of the running order of speakers and a brief description of what was said is extremely helpful at this point. (For a full discussion of transcribing interviews, see Chapter 4.)

Read the transcribed material over several times as a means of becoming familiar with the discussion. One relatively straightforward strategy for proceeding draws from the questions or themes that focused the discussion. Write each question or theme on the top of a separate sheet of paper and then on each sheet list the relevant points made. Finally take a note of key quotes that might be used in written material (Bertrand, Brown and Ward 1992). This approach works well when the discussion did not deviate widely from the questions or themes set by the researcher, or when comparisons are to be made between focus groups (Bertrand, Brown and Ward 1992). For example in a research project comparing the land management strategies for dealing with salinity preferred by farmers, policy-makers and researchers, the sheets with the responses of the different groups to each question or theme might be compared easily.

When the purpose of the research project is to identify key themes or processes associated with a particular issue or topic it may be more appropriate to use margin coding (Bertrand, Brown and Ward 1992). To do this, read through the transcripts, identify key themes or categories, and devise a simple colour, number, letter or symbol-based coding system to represent the themes or categories. The transcripts should then be reread; words, sentences, and paragraphs related to each category or theme are highlighted by writing the appropriate code in the margin. Once transcripts have been coded, a cut and paste technique—completed either on a computer or manually—can be used to group the discussion related to each theme or process (see Chapters 4 and 8 for more information on this). Always keep a copy of transcripts intact for future reference. When analysing focus group discussions of social and economic change in regional communities, Gibson, Cameron and Veno (1999) distinguished between those understandings of change that seemed to dominate the discussions and other more marginal understandings. A print-out of the transcripts was grouped according to these two themes and used as the basis for discussion in the final written document. Sherraden (n.d.) suggests a variation of this thematic analysis. He develops a list of key words and, in a word-processing package, types two or more key words beside each comment. Using the search function of the word processor, it is then possible to locate related points of discussion. Computer programs specifically designed for qualitative analysis, like *NUD*IST*, can also be used, and are particularly helpful when there is a large amount of transcribed material to be analysed (see Chapter 8 for a discussion of this).

Being able to find material quickly is an important consideration as analysis and writing rarely proceed in a linear fashion. During the writing process new insights unfold (see Chapter 9) and frequently you may find it necessary to return to the original transcripts to refine and reformulate ideas. Sometimes it will be necessary to listen to and make additional transcriptions of sections of tape.

When reporting on focus group research, present your results only in terms of the discussion within the groups. As noted earlier, focus groups do not produce findings that can be generalised to a wider population. Focus group results are also expressed in impressionistic rather than numerical terms. In place of precise numbers or percentages, the general trends or strength of feeling about an issue are typically given. As Ward, Bertrand and Brown (1991, p. 271) have noted, focus group reports are 'replete with statements such as "many participants mentioned …," "two distinct positions were observed among the participants", and "almost no one had ever …"'. Reporting on their study into people's responses to emergency procedures, Zeigler, Brunn and Johnson (1996), for example, noted that the people in their focus groups generally responded with either compliant behaviour or under-reaction. Zeigler, Brunn and Johnson then used direct quotes to illustrate the different ways that the responses were expressed (see also the ways that Jackson and Holbrook [1995], Jarrett [1994], and Myers and Macnaghten [1998] discuss focus group findings).

In some projects it might not be the general trends but the ambiguous or contradictory remarks and points of discussion that the researcher particularly wants to explore. The development and presentation of an argument may refer not only to *what* was talked about, but the *way* it was talked about in the group setting. This became an significant aspect of the focus group research conducted by Gibson, Cameron and Veno (1999, p. 29):

> The stories of success and hope that emerged when the discussion was shifted onto the terrain of community strengths and innovations were numerous. They came stumbling out in a disorganised manner suggesting that these stories were not readily nor often told. In the face of dominant narratives of economic change perhaps such stories are positioned as less important or effective. It is clear that there is a lack of a language to talk about this understanding of community capacity; yet, as we will argue, this understanding has the potential to contribute to the ability of a region to deal effectively and innovatively with the consequences of social and economic change.

One important element in the process of writing up (or 'writing-in' as Berg and Mansvelt call it in Chapter 9) is to find a balance between direct quotes and your summary and interpretation of the discussion. When too many quotes are included the material can seem repetitive or chaotic. Too few quotes, on the other hand, can mean that the vitality of the interaction between participants is lost to the reader. Morgan (1997, p. 64) recommends that the researcher should aim to connect the reader and the original participants through 'well-chosen' quotations (see Chapter 4 for more information on this).

CONCLUSION

Focus groups demand careful preparation on the part of the researcher. The selection and recruitment of participants; the composition, size, and number of groups; and the questions and topics to be explored are all key points to consider in the planning stage. Even the apparently mundane details of appropriateness of venue, provision of refreshments and quality of audio equipment are critical to the success of focus groups. A well-prepared researcher will also give thought beforehand to the process of facilitation, including the points to cover in the introduction to the session; the wording of key questions, topics and phrases; the probes and prompts that might be useful to explore further a theme or topic; and strategies for drawing out different points of view, keeping the discussion on track, and dealing with more talkative and quiet members of the group. As soon as possible after the focus group start the process of analysis, beginning with the transcription of the audio-tapes and followed by the reading and rereading of transcripts, the summarising of main points and the identification of central themes.

Although they require careful planning beforehand and a great deal of reflection afterwards, focus groups are an exciting and invaluable research tool for geographers to use. Participants almost invariably enjoy interacting with each other, and the discussion can generate insights and understandings that are new to both participants and researchers. The interactive element makes focus groups ideally suited to exploring the nuances and complexities of people-place relationships, whether the research has a primarily data gathering function or is more concerned with the collective practice of knowledge production.

KEY TERMS

disclosure
discourse
facilitator
margin-coding
moderator
probe
purposeful sampling
recruitment
transcription

REVIEW QUESTIONS

1. Find a research project from a recent issue of a geographical journal that you think could have been conducted using focus groups. Why do you think focus groups would be appropriate? Discuss the participants you would select, the composition of the focus groups, the size and number of groups you would use, the questions you would ask or themes you would use, and strategies for recruiting participants.
2. The University of Pacifica is planning to upgrade its indoor sports facilities. Your research company has been commissioned by the University to conduct a focus group study on the sorts of changes that students think are most important. Discuss your research plan, including the composition, size and number of focus groups; the process of selecting and recruiting participants; and the questions or topics you would use. Make sure you provide a rationale for each of your research decisions.
3. In a few days time you will be facilitating a series of focus groups. Describe the final steps you would take to prepare for the groups. What are some of the issues you anticipate might arise when conducting the groups? Discuss the strategies you would use to manage these. You might want to consider the following issues: too much agreement between participants; over-disclosure of personal information; groups that are overly talkative.

SUGGESTED READING

Dick, B. 1997, *Structured Focus Groups* [on line], available at http://www.scu.edu.au/schools/sawd/arr/focus.html

Lewis, M. 1995, *Focus Group Interviews in Qualitative Research: A Review of the Literature* [on line], available at http://www.scu.edu.au/schools/sawd/reader/rlewis.html

Morgan, D.L. 1997, *Focus Groups as Qualitative Research*, 2nd edn, Sage, Thousand Oaks.

Sherraden, M. n.d., *How to do Focus Groups* [on line], available at http://www.gwbssw.wustl.edu/fgroups/fghowto.html

Stewart, D.W. and Shamdasani, P.N. 1990, *Focus Groups: Theory and Practice*, Sage, Newbury Park.

6

Being There: Research through Observing and Participating

Robin A. Kearns

CONTENTS

CHAPTER OVERVIEW

'... for a discipline often preoccupied with the visual, it seems ... [geography] has not studied the practices of seeing rigorously enough' (Crang 1997b, p. 371)

• • •

'Seeing is believing', as the saying goes. While visual observation is a key to many types of research, there is more to observation than simply seeing: it also involves touching, smelling and hearing the environment, and making implicit or explicit comparisons with previous experience (Rodaway 1994). Further, seeing implies a vantage point, a place—both social and geographical—at which we position ourselves to observe and be part of the world (Jackson 1993). What we observe from this place is influenced by whether we are regarded by others as an 'insider' (i.e., one who belongs), an 'outsider' (i.e., one who does not belong and is 'out of place'), or someone in between. The goal of this chapter is to reflect on the reasons why observation is fundamental to geographical research, and to consider critically what it means to observe. It explores various positions the researcher can adopt vis-à-vis the observed—from the viewing of secondary materials such as photographs to participating in the life of a community. Two key contentions are:

1. *that observation has tended to be an assumed, and consequently undervalued, practice in geographic research; and*
2. *that ultimately all observation is participant observation.*

In the chapter, I argue that observation has been taken for granted as something that occurs 'naturally', and therefore does not require the attention granted to more technical aspects of our methodological repertoire such as questionnaire design or survey sampling. With critical reflection, however, observation can be transformed into a self-conscious, effective and ethically sound practice.

PURPOSES OF OBSERVATION

Among the definitions of **observation** in the *Oxford English Dictionary* is 'accurate watching and noting of phenomena as they occur', implying that observation has an unconstrained quality. To regard observation as random or haphazard would be a mistake, however, for we never observe everything there is to be seen. Observation is the outcome of active choice, rather than mere exposure. Our choice—whether conscious or unconscious—of first, what to see, and second, how to see it, means we always have an active role in the observation process. Following Mike Crang, I wish to argue for observation as a way of 'taking part in the world, not just representing it' (Crang 1997b, p. 360).

There is a range of purposes for observation in social scientific research and these can be summarised by three words conveniently beginning with 'c': counting, complementing, and contextualising. The first purpose—counting—refers to an enumerative function for observation. For example we might accumulate observations of pedestrians passing various points in a shopping mall or airport in order to establish daily rhythms of activity within these places. Research in the time-geographic tradition has used this approach to chart the ebb and flow of spatio-temporal activity (Shapcott and Steadman 1978; Walmsley and Lewis 1984). Under this observational rationale, other elements of the immediate setting are (at least temporarily) ignored while the focal activity occurs. The resulting numerical data is then easily subjected to graphical representation or statistical analysis. This is an approach to observation which may be useful for establishing trends, but is ultimately too reductionist to develop comprehensive understandings of place.

A second purpose of observation is providing *complementary* evidence. The rationale here is to gather additional descriptive information before, during, or after other more structured forms of data collection. The intent is to gain added value from time 'in the field' and to provide a descriptive complement to more controlled and formalised methods such as interviewing. Complementary observation might involve spending time in a neighbourhood after completing a household survey and taking notes on the appearance of houses, the types of cars and the upkeep of gardens. I used this approach in research on housing problems in Auckland and Christchurch. Interviewers were instructed to observe the dwellings of those they interviewed. Their notes added to (and often contrasted with) what people said about their own dwellings (Kearns, Smith and Abbott 1991). Such information complements the aggregated data gathered by more structured means and assists in interpreting the experiences in places.

The third purpose of observation might be called *contextual* understanding. Here the goal is to construct an in-depth interpretation of a particular time and place through direct experience. To achieve this understanding the researcher immerses herself/himself in the socio-temporal context of interest and uses first-hand observations as the prime source of data. In this situation, the observer is very much a participant.

It should be noted at the outset that these purposes are not mutually exclusive. As I will later show by way of examples from research, one can approach observation with mixed purposes, seeking, for instance, to both enumerate and understand context during a period in the field (Kearns 1991a).

TYPES OF OBSERVATION

Some social scientists identify two types of observation: *controlled* and *uncontrolled*. **Controlled observation** 'is typified by clear and explicit decisions on what, how, and when to observe' (Frankfort-Nachmaias and Nachmaias 1992, p. 206). This style of observation is associated with natural science and its experimental approach to research, which has been imported into physical geography. Thus, a geographer might set up an electronic stream gauge to gather data at periodic intervals throughout a flood, resulting in a series of so-called 'observations'. However, such data are collected remotely without the aid of human senses. Indeed, the human eye may have only been involved in the secondary sense of observing the gauge, subsequent to the actual data collection. This recognition that controlled observations can be made through mechanical means adds weight to the common perception that such research is rigorous and easily replicated. However, controlled observation is also limiting in terms of the sensory and experiential input admissible as 'findings'. Human geographers are unlikely to employ such controlled methods of observation, except perhaps in the earlier example of pedestrian counts.

In two respects, so-called 'controlled observation' is necessarily limiting. First, there is an imposed focus on particular elements of the known world; and second, it is only *directly* observable aspects that are of interest. Thus, imputed characteristics of place and the feelings of residents may be out of range. Most of the observation conducted by contemporary social and cultural geographers could be described as 'uncontrolled'. Such observation is certainly directed by goals and ethical considerations, but is not controlled in the sense of being restricted to noting prescribed phenomena. Therefore, although observation refers literally to that which is seen, in social science it may involve more than just seeing. Most obviously, observation also includes listening, a critical aspect of several approaches in human geography, including interviewing and participant observation. Effective listening can assist visual observation by both confirming the place of the researcher as a participant (Kearns 1991b) and by attuning oneself to 'soundscapes' and the aural aspects of social settings (Smith 1994).

One might argue that all research involves observation, or at least is made up of a series of observations. Thus, in a social setting, we can 'observe' the population by employing questionnaires through which the researcher establishes the frequency of certain variables (for example, occupation, sex). The activity of conducting a questionnaire survey invariably places the researcher in the position of an 'outsider', marked as 'other' by purpose, if not appearance and demeanour (for example, clothing, age, ethnicity, or type of language used). Identifying a sample and evaluating the responses to questions involve concerns related to the goal of generalisability to broader populations. But the cost of these concerns is that only a restricted subset of social phenomena are deemed to be of interest. These phenomena can easily be isolated from their context, unaccompanied by less directly observable values, intentions and feelings. Qualitative approaches allow the consideration of human experience, potentially suspending traditional concerns about researcher bias and recognising instead the relationship between the researcher and the people and places he or she seeks to study.

A further distinction can be made between **primary** and **secondary observation** in research. The former activity would have us adopt the position of participants in, and interpreters of, human activity, whereas as secondary observers we are interpreters of the observations of others. Examples of secondary observation are analyses of picture postcards available to tourists (Crang 1996) and the photography of one 'racialised' group by a representative of another (Jackson 1992). In this chapter, I will not dwell on this use of secondary materials (which are dealt with fully in Chapter 7), but rather note the all-pervasive nature of participant observation. For, in one sense, a solitary researcher observing images of place is an active participant inasmuch as they are co-creating meaning through bringing their own perspectives and life experiences to analysis and interpretation.

PARTICIPANT OBSERVATION

Participant observation is most closely associated with social anthropology (Sanjek 1990; Srivinas, Shah and Ramaswamy 1979) and, significantly, its profile among the repertoire of approaches used by human geographers was heightened by Peter Jackson (1983) who had studied both geography and social anthropology. The approach has been adopted and adapted by geographers seeking to understand more fully the meanings of place and the contexts of everyday life. Examples include a number of now 'classic' studies within the 'humanistic' tradition such as David Ley's (1974) work on 'Monroe', a neighbourhood in inner-city Philadelphia, John Western's (1981) *Outcast Cape Town*, and Graham Rowles' (1978) research on sense of place among elderly persons. Like many social geographers, these writers talked to 'locals' in the course of their research, but it was the depth of their involvement in a community, their recurrent contact with people, and their relatively unstructured social interactions that stood out in their work. While many of their contemporaries were solely interested in *perception* or *behaviour*, these geographers were concerned with *experience*.

Developing a geography of everyday experience requires us to move beyond reliance on formalised interactions such as occur in interviews. As Mel Evans (1988, p. 203) remarks, 'although an interview situation is still a social situation .... it is a world apart from everyday life'. Evans is suggesting that no matter how much we are able to put people at ease before and during an interview, its structured format often removes the researcher from the 'flow' of everyday life in both time and space. In other words, an interview ordinarily has an anticipated length and occurs in a mutually agreeable place often set apart from other social interactions. In contrast, participant observation is concerned with developing understanding through being part of the spontaneity of everyday interactions.

There is a consensus among commentators that, although definable, participant observation is difficult to describe systematically. They have remarked on its 'elusive nature' (Alder and Alder 1994), its breadth (Bryman 1984) and the fact that it remains 'ill-defined and .... tainted with mysticism' (Evans 1988, p. 197). Part of the 'mysticism' Evans refers to stems from the fact that there are few systematic outlines of the method, and students who have been curious about the approach have commonly been referred to 'classic' studies such as William Whyte's *Street Corner Society* (1957) for models. One explanation for this tendency to refer to examples rather than to offer step-by-step guidelines is that every participant-observation situation is unique. Another reason offered by Evans (1988, p. 197) is that the success of the approach depends less on the strict application of rules and more upon 'introspection on the part of the researcher with respect to his or her relationship to what is to be (and is being) researched'.

While introspection or reflection on what we see and experience is important, it is surely guidance in the *act* of observing that is needed. Jackson draws on Kluckhohn (1940) for a concise description of participant observation, defining it as 'conscious and systematic sharing, in so far as circumstances permit, in the life activities and, on occasion, in the interests .... of a group of persons' (1983, p. 39). It is therefore the systematic and intentional character of observations that contrasts the activities of a participant observer with those of routine participants in daily life (Spradley 1980). To generalise, participant observation for a geographer involves strategically placing oneself in situations in which systematic understandings of place are most likely to arise.

One rationale for participant observation is a recognition that the mere presence of a researcher potentially alters the behaviour or the dispositions of those being observed. This contention is illustrated in a well-known 'Far Side' cartoon by Gary Larson. In the drawing, two men wearing pith helmets are approaching a thatched hut. The 'primitive' occupants, stereotypically adorned with bones through their noses and wearing head-dresses, are exclaiming 'Anthropologists! Anthropologists!' as they rush to hide their television set. The cartoon ironically suggests that scholarly explorers do not expect 'primitives' to have technology (and that primitives might want to live up to this stereotype). In terms of research methods, the cartoon's message is that undisguised observation is bound to alter behaviour. This point is a reminder that conscious participation in the social processes being observed increases the potential for more 'natural' interactions and responses to occur.

Use of the term 'observation' in highly controlled scientific research perhaps tempts us to think too easily in terms of a simple dichotomy between participant and non-participant. According to Atkinson and Hammersley, this is an unhelpful distinction 'not least because it seems to imply that the non-participant observer plays no recognised role at all' (1984, p. 248). Indeed, as the Far Side cartoon succinctly illustrates, there is really no such thing as a non-participant in a social situation: even those who believe that they are present but not participating in a research context often unwittingly alter the research setting. To move beyond this false binary construct of participant/non-participant, Gold (1958) suggests that there is a range of four possible research roles:

1 **complete observer** (for example, a psychologist watching a child through a one-way mirror);
2 **observer-as-participant** (for example, a newcomer to a sport being part of the crowd—see Latimer 1998);
3 **participant-as-observer** (for example, seeking to understand social change in one's own locality—see Ponga 1998);
4 **complete participation** (for example, living in a rural settlement to understand meanings of sustainability—see Scott et al. 1997).

While it is difficult to imagine how being a complete observer might be incorporated into geography, an example could be discrete surveillance in a shopping mall, perhaps with access to closed-circuit television (CCTV). As the references above indicate, recent work by geographers can be easily categorised according to the remaining three roles. Each represents a form of participant observation, and the difference is essentially a matter of the *degree* of participation involved. This division implies a continuum of involvement for the observing researcher from detachment to engagement. However, whatever one's **positioning** *vis-à-vis* 'the observed' it is important to acknowledge that the act of observation is imbued with power dynamics.

POWER, KNOWLEDGE AND OBSERVATION

Before considering the stages of observation in the field in greater detail, it is important to reflect on aspects of the process itself. As emphasised earlier, observation involves participating, both socially and spatially. We cannot usually observe directly without being present, and bodily presence brings with it personal characteristics such as 'race', sex and age. Belonging to dominant groups in society can mean we potentially carry with us the power dynamics linked to such affiliations. Being a white, adult male, for instance, will invariably create challenges to being a participant in a group whose members do not share those characteristics, such as a new mothers' support group. In other words, our difference in terms of key markers of societal power (or lack thereof) contributes to our (in)ability to be 'insiders' and participants in the quest to understand place (matters of 'insider' and 'outsider' status are also discussed in Chapter 2).

More subtle challenges may be generated by our level of education and affiliation to universities. In undertaking university-based research, we can carry institutional dynamics into our acts of observing through subtle forms of social control. Social control can occur through the ways in which one group (the powerful) are able to maintain watch over members of another (the disempowered). Perhaps the most memorable image of this dynamic is Bentham's '**panopticon**', a circular prison designed to maximise the ability of warders to see into every cell and to watch prisoners (Foucault 1977a). Prisoners do not know exactly when they are being observed but learn to act as if they were always being watched. Foucault sees this surveillance as a form of disciplinary power which is enforced through the layout of the built environment rather than through the exertion of force *per se*. The key to the resulting institutional dynamic is the knowledge on the part of prisoners that they *could* be observed at any time. This knowledge, according to Foucault, results in self-surveillance and self-discipline.

Foucault's ideas have implications for observation as a research approach. The 'surveillance' to which he refers to is a very visual and disembodied form of observation. Gillian Rose (1993) has linked this to the traditional geographical activity of fieldwork. To Rose and other feminists, geography has been an excessively observational discipline characterised by an implicit 'masculine gaze'. The key point is that observation can be a power-laden process deployed within institutional practices. As we are based in, and representative of, academic institutions, it is imperative that we be aware of the ways in which others' behaviour may be modified by our presence (Dyck and Kearns 1995).

A challenge posed by feminist geographers and anthropologists is to see fieldwork as a gendered activity (Bell, Caplan and Karim 1993; Nast 1994; Rose 1993). Their argument is that in being participant observers, we unavoidably incorporate our gendered selves into the arena of observation. An extract from the fieldnotes of Louisa Kivell (1995) illustrates how observation in human geography is far from a simple matter of unidirectional watching; rather it involves interactions that are potentially laden with sexual energy. The resulting situations may shed light on the gendered constitution of the field site itself (Box 6.1).

Box 6.1: 'And black underwear too': the gendered constitution of a field site

Me. Walking onto an orchard looking for the manager with whom I had an appointment to discuss employment relations in his operation. Me. Dressed in my 'orchard clothes', a reflection of my expectation that he would have been out working in his 'orchard clothes' too. My orchard clothes include: dark shorts suitable for ladders, dirt and other eventualities; a plain green T-shirt without any brand labels or other logos; a jumper tied around my waist because I know it can get cold down the apple rows; socks and cross-trainers .... Anything else? Oh yes, a band tying my hair out of the way, a watch and a medic alert bracelet, mascara, suntan lotion and deodorant, lip balm, a pen and paper in my pocket and maybe a linger of this morning's perfume. Stop.

A white male wearing his 'orchard clothes' approaches me on a tractor and stops. He smiles, then laughs out a question-come-statement: *'You aren't here looking for a picking job are you?'*

For a moment I am incapable.

I reply. *'As it happens, no. I have an appointment to see Mr Sky but I'm interested to hear what makes you think I wouldn't be picker material.'*

'Oh well .... nothing really .... its just that um .... in two years I've never seen a girl like you on an orchard before.'

What is a girl like me?

White, 21, pretty, on my own.

My education status, my socio-economic background, my place of origin, most of the other 'girl like me' things are not immediately visible.

What pale concept of a girl like me did he dispatch with his laugh? (Kivell 1995, p. 49).

STAGES OF PARTICIPANT OBSERVATION

Participant observation is far from haphazard. There are some commonly recognised stages through which the process moves, from choice of research site through to presentation of results. I will review each of these, using as a case example my research in the Hokianga district of Northland, New Zealand. In this work, my objective has been to understand the influence of sense of place in shaping the social meanings of the local health system. In the course of this research, I have adopted three of the four roles identified by Gold (1958): observer-as-participant, participant-as-observer, and complete participant (see Kearns 1991b, 1997).

Choice of setting

Choosing a setting for participant observation might be dictated by the goals of a larger research project, in which case the setting may be unfamiliar to the researcher (for example, Scott et al. 1997). In such a case, there will be a need for a good deal of background research on the community in question as well as reconnaissance visits before a period of immersion into community life. However, it is likely that student researchers will choose to study settings at least partially familiar to them. But familiarity can bring pitfalls. There is a danger that the researcher is over-familiar with the community, with the result that there is 'too much participation at the expense of observation' (Evans 1988, p. 205). Indeed some argue that it is just as challenging doing fieldwork in one's own society as in an 'alien' one (Srinivas, Shah and Ramaswamy 1979).

What then might be the ideal? Possibly the best balance to strive for in choosing a research setting is a mid-point between the 'insider' and 'outsider' statuses discussed earlier. The conventionally recommended stance is that of stranger, but this is not necessarily a position in which one simply does not belong. Rather, it is one in which the researcher's status is what Evans (1988) terms 'marginal' in relation to the community. By 'marginal' I take Evans to mean (socially) on the edge of a community or group. Thus, although cheering on the terraces at Eden Park (a sports ground in Auckland), Bill Latimer (1998) was marginal because he had never been to a rugby game before. Coming from northern England, the game of his culture and class was soccer. He was thus able to be a critical, only partially involved, observer of the place of rugby in Auckland (see Box 6.2).

Box 6.2: At the game: an observer-as-participant

Saturday March 21st, walking alone to Eden Park I was struck by the sea of decorative clothing and painted faces—blue and white for Auckland; red and black for Canterbury. One man had made himself a hat out of DB Export beer coasters. Once inside the ground I steered away from the loud 'yobs' who had positioned themselves behind the goal posts. I didn't choose a good spot. I felt uneasy as I took my seat between two sets of monotonic droning young men (wailing 'AAAAUUUUAUCKLAAAAAAND!!!' like Tarzans of the terrace).

An eerie silence accompanied the match, interrupted only by the mo(ro)notonic drone of those seated around me. As I was about to discover, this was the calm before the storm. The Blues scored. Before I was able to offer a congratulatory clap of hands Eden Park erupted. Music blasted onto the terraces and was taken as the cue to scream, dance and throw arms, and their contents, into the air. As the music continued to pump out, cheerleaders thrusted and gyrated suggestively to the very deep, sexy, sensuous base beat. I was soaked with beer and hit about the body with plastic cups and other missiles. I was also very unimpressed. Each time Auckland scored, a young guy in front of me would stand up and say to his friend, 'did you feeeel that?'. He would then embrace his mate in a hand wrestling fashion which meant that their forearm muscles bulged, and both would look each other in the eye, faces inches apart, and grunt 'AAAAUUUUUUCKLAAAAND!!!' (Latimer 1998, p. 91).

Access

Gaining entry to social settings and places is potentially a fundamental challenge. A crucial issue is identifying key individuals who can act as gatekeepers, facilitating opportunities to interact with others in the chosen research site. There are some settings into which one can simply walk and take on the role of participant. Investigating the way in which a public shopping mall is used is a good example. Being commonly used and perceived as a public place, there are fewer permissions to be sought than in less 'public' settings such as health clinics. And nor is there a problem with 'blending in', given the diversity of users. One can observe through simply participating in the mall's functions: shopping, resting on the seats provided, or using the food court. But gaining access is potentially more challenging when the place is smaller in scale or less public in character.

Gaining access may well be more straightforward if one has a known role. Even being 'the visiting student' in a workplace may give one a role in a way that just being an anonymous visitor or stranger would not. However, there are often no convenient roles in hospitals or factories, so once ethical approval is gained (see the discussion in Chapter 2), perhaps there is good reason to resist being typecast into any role except that of outsider (see Kearns 1997). This ambiguous position worked for me in the health clinics of the Hokianga. I could not legitimately pass as either a health professional or a patient as local residents knew each other too well, so being a visitor reading the newspaper helped me be reasonably inconspicuous for a short period (see Box 6.3).

Box 6.3: Gaining access to a field site

My research in, on and with Hokianga people began with a period of fieldwork during October 1988. My wife Pat had the opportunity of being a medical *locum tenens* in the district and I was also keen to go north. Once in Hokianga, having a spouse with a temporary part in the health-care system gave me some legitimacy in seeking to gain approval to undertake some research at the hospital and community clinics. The medical director as gatekeeper to the health system knew who I was, and as I was already there the formality of letters and telephone conversations could largely be circumvented. I hastily devised a research plan that involved spending time in the waiting areas of each clinic, where I observed social dynamics with the goal of understanding the social function of these places.

Field relations

The role that the researcher adopts within an observed setting (for example, complete participant or participant-as-observer) will define the character of the relations she or he generates. The idea of impression management is critical here, for the impact you make on those encountered will determine, to a large extent, the ease with which they will interact with you and incorporate you into their place. Embedded within the word *incorporate* is 'corpus', the Latin word for body. My purpose in discussing incorporation is to stress the idea of the researcher's embodiment, and to recognise that as researchers we take more than our intentions and notebooks into any situation: we take our bodies also. The way we clothe ourselves, for instance, can be a key marker of who we are, or who we wish to be seen as, in the field. By way of example, while researching the inner-city experiences of psychiatric patients, I chose to wear older clothes to drop-in centres in order to minimise being regarded as yet another health professional or social worker intruding on patients' lives (Kearns 1987). Being a student at the time, it was easy for me to find place-appropriate clothes! But a troubling question is whether I could have 'blended in' if I had been attempting to study the more elite social relations of place (for example, the dynamics of a lawyers' convention, or a restaurant frequented by politicians). At the most fundamental level, I would not have had the right clothes to wear! It is generally easier to dress 'down' than 'up' and this perhaps explains in small part why participant observation is more often used in studies of people less powerful than researchers themselves.

Our ability to relate to others in the field depends not just on appearance, but also on the level and type of activity undertaken. To extend the above example, passively 'hanging out' in a psychiatric drop-in centre is one thing, but feigning ability in more expert pursuits is another. This need to pass as an *active* participant is reflected in Phil Crang's (1996) research on the workplace geographies within a Mexican theme restaurant in England. Here the researcher sought to understand the dynamics of performance in the restaurant by being employed as a member of the waiting staff and becoming part of the daily routines of the place.

A further point is that although concern for appearance and clothing is appropriate, it potentially reinforces geographers' fixations with the visual (see Rose 1993). Hester Parr (1998) describes how, while researching the geographical experiences of people with mental illness in Nottingham, she became conscious of non-visual aspects of her 'otherness' such as smell: her perfumed deodorant served to set her apart and inhibit free interaction. Parr's reflection serves to remind us that senses such as smell add to the character of place (Porteous 1985) and merit consideration for the way they can mark as 'other' the bodiliness of the researcher (see Rodaway 1994).

A related point influencing field relations is that codes of behaviour are attached to different settings. Hester Parr (1988) discusses how she suddenly established a rapport with Bob, an acutely schizophrenic man with whom communication was difficult. The setting for this 'break-through' was a city park, neutral ground which neither the researcher nor the informant regarded as home. This site at which social boundaries were blurred contrasted with the drop-in centre where roles were more established and routine. The lesson is that research relations may be enabled or constrained by the (often unspoken) ways in which social space is codified and regulated. Successful participant observation thus involves not only the (temporary) occupation of unfamiliar places, but also the adoption of alternative ways of using time. Box 6.4 continues an account of my Hokianga research by describing my use of time in community clinics.

Box 6.4: Field relations in an observational study

At some of the ten hospital and community clinics I blended into the small crowd of clinic attendees, inconspicuously observing events under the guise of reading the newspaper. At others, however, I was clearly a Pakeha ('white') visitor, as all present were both local and Maori and seemed to know each other well. It was immediately evident that going to the doctor involved more than medical interactions. Rather, the occasion frequently provided an opportunity for locals to tell stories and reflect not only on their own well-being, but also on that of their families and friends. From this participant observation in the clinic waiting areas, I noted that the most frequent conversation category was community concerns. Comments on the deleterious impacts of the restructuring of public services were frequently expressed. Residents adopted a relaxed approach to clinic attendance and their use of waiting areas. Patients were observed arriving well before their appointment time (sometimes in the company of others who had no intention of consulting the doctor or nurse) and lingering afterwards 'having a yarn'. These observations led me to interpret the clinics as de facto community centres, analogous to the village market in other countries. The difference, however, was that the place (the clinics) and the time (clinic day) prompted a considerable amount of conversation that explicitly centred on health concerns (adapted from Kearns 1991b, p. 525).

Talking and listening

We cannot blend in as researchers unless we participate in the social relations we are seeking to understand. Listening ought to precede talking, in order to become attuned to what matters in a particular time, place and social setting (Kearns 1991a) (see Box 6.5). Asking questions can be manipulative, and simply conversing about what otherwise might seem obvious may result in a less threatening entry into the social relations of place. The how and where of talking and listening are also crucially important. For research with children, for instance, adopting their level—both physically and in terms of style of language—may be the key to successful observation. Sarah Gregory (1998) found that playing with children on the floor was a useful precursor to asking them about their experiences using consumption goods. In other words, observation is least conspicuous when one is interacting most naturally with the research subjects. The lesson is that as researchers we are our own most crucial tool and 'it is the fact of participation, of being part of a collective contract, which creates the data' (Evans 1998, p. 209).

Box 6.5: Talking and listening: the embodied observer

My embodiment as researcher was central to the construction of this knowledge. Within the waiting area I had to position myself in such a way as to neither be threatening (and thus inhibit conversation by gazing at others) nor be overly welcoming of conversational engagement (hence my 'hiding' behind a newspaper). Such choreography was bound to break down, and did. On one occasion two locals offered to help with the crossword puzzle I was half-heartedly completing, and on another, a *kuia* (female elder) entered, kissed and welcomed all present to 'her' clinic, including myself. My corporeality within the observed arena of social interaction thus rendered binary constructs of researcher/researched and subject/object thoroughly permeable. In this *kuia's* clinic, my conceptions of being an 'autonomous self' were dissected and (re)embodied within their rightful web of socio-cultural relations (Kearns 1997, p. 5).

Recording data

A clipboard or tape recorder are the standard means of recording information in other qualitative approaches involving face-to-face communication, such as interviews. However, because these tools would be disruptive to the flow of conversation or interaction, a participant observer can rarely use them. Rather there is a greater reliance on recollection and a necessity to work on detailed note-taking after a period of field encounters. At the end of any day or session of observation, one is likely to feel tired and not inclined to take out a pen, or go to a computer to record reflections. However, developing a discipline for such 'homework' is a key part of field observation: notes are invaluable sources of data, and prompts to further reflection (Scott et al. 1997). **Fieldnotes** become a personal text for the researcher to refer to and analyse. They represent the process of transforming observed interaction into written communication (see the discussion in Chapter 9 on relationships between writing and research). Jean Jackson (1990, p. 7) describes fieldnotes creatively as 'ideas that are marinating'. Preliminary fieldnotes may be taken on any materials at hand, such as the margins of a newspaper (Kearns 1991b). However, back at 'home base', annotations are now almost invariably entered into a computer, and just as a fear of earlier researchers was destruction of paper notes by fire (Sanjek 1990), a contemporary concern is loss of electronic data. Keeping back-up files and/or print-outs of your field data is thus a crucial precaution.

Analysis and presentation

Analysing the results of observation will vary according to the purposes for which it was undertaken. Observations that have involved *counting*, or carefully recording each instance of some phenomenon, lend themselves to tables that enumerate occurrences and express data as frequencies or percentages. Observation that has been *complementary* to a more explicitly structured research design commonly leads to the presentation of quotations or descriptions that assist in interpreting findings derived from other sources. In research into the impact of logging trucks on Hokianga roads and people, we complemented interviews with community members and analyses of secondary data, such as log-harvest projections, with a first-hand narrative of the experience of driving along selected stretches of road (Collins and Kearns 1998).

When observation is embedded in an attempt to *reach contextual understanding*, a considerable volume of text will typically be accumulated, and strategies for the storage, classification, and analysis of information will need to be carefully considered. A number of useful texts are beginning to appear (for example, Bryman and Burgess 1994). This scale of observational engagement also suggests the use of computing software such as *NUD*IST*, now employed by human geographers (see Chapter 8).

Ethical obligations

What are our obligations to those whom we have observed? Clearly if observation has been fleeting and devoid of personal contact (for example, watching pedestrian behaviour) any return of research results to those individuals observed would be impractical and perhaps unnecessary. Indeed the 'observed' in this example are only nominally 'participants' in the research. However, where observation 'involves involvement', it might be argued that there is an ethical imperative to maintain contact. This imperative may be formalised by the requirements of university ethics committees, but the stronger influence should surely come from the researcher, especially if pre-existing relationships have been developed (or reactivated) through the research process. As Maria Ponga noted in her study of the restoration of a *marae* (meeting house) in her home community, 'there is a social responsibility to carry on the ties. I have tried to keep in touch and send letters thanking people for the information they have given me and keep them updated on my progress. As we are all *whanau* (extended family), this has also had an added dimension of maintaining the family ties' (1998, p. 54).

Such obligations may be taken for granted when pre-existing ties are involved, but require careful consideration when social situations are entered into, or generated, for the sake of research. It is generally agreed that cross-cultural fieldwork is particularly problematic. This is for two reasons that may be linked to the 'field' metaphor: first, the researcher is potentially venturing onto another's turf; and second, because fieldwork involves a researcher working in a field of knowledge, there is the risk of eliding local understandings and priorities, and of (possibly unintended) one-way traffic of knowledge from the field (periphery) to the academy (centre) (Kearns 1997). Box 6.6 draws on these ideas to complete the series of examples from my Hokianga research. In such situations, the development of '**culturally safe**' research practice (Dyck and Kearns 1995) is important. Such practice recognises the ways in which collective histories of power relations may affect individual research encounters. It also stresses the need for appropriate translation of materials into everyday language, and the return of knowledge to the communities that provided or generated it (Kearns 1997). (See also Chapter 3.)

Box 6.6: Completing the circle: cultures, theory and practice

While on overseas sabbatical leave in 1995, I had the opportunity of recounting the story of place and health in Hokianga, and connecting its plot lines to the coordinates of other struggles for identity and turf. Drawing on Feinsilver's (1993) Cuban experience, for instance, I could identify the community action as a 'narrative of struggle' in which there is a greater symbolism to health politics than just a defence of local services. However, I was left searching for a rationale for the place of the researcher in the narrative of struggle.

To impose theory (upon observation) without reference to the community would be as foolish as the unfettered importation of exotic wildlife or viruses into New Zealand. Clearly, the use of theory must be regulated. My own form of self-regulation has been to return draft papers to people involved with Hokianga's health trust for comment. In one such draft I had interpreted their struggle as a 'postmodern politics of resistance' (Kearns 1998). The returned manuscript was annotated with the comment from a health trust worker that my words sounded like 'undigested theory'. On reflection, this was a fair assessment. I had, perhaps unwittingly, used theory imported from recent visits to conferences in Los Angeles and Chicago. It was time to return, to digest theory together and reconnect with the source of the story. Such returns are made easier through adoptive *whanau* (extended family) relationships in which research is, at times, indistinguishable from the *aroha* (love, affection) and *korero* (purposeful talk) among friends. Hokianga has taught me about being a bicultural geographer, a role that requires respecting the rituals we can never fully enter, while reforming the rituals of our own research (Kearns 1997, p. 6).

REFLECTING ON THE METHOD

As observers, our goal should be to achieve seeing that is, as the saying goes, believing. However, this chapter has argued that believable observation is the outcome of more than simply seeing; it requires cognisance of the full sensory experience of being in place.

Though a well-established approach—especially within anthropology—participant observation can be reinterpreted as a quintessentially postmodern activity. This is because its goal is to acknowledge difference and, through immersion in a situation, to 'become the other', however provisionally. If the questionnaire is the tool for survey researchers, and the tape recorder for key informant interviewers, the researcher herself or himself is the tool for participant observation. The question remaining is: how do we know if the fruits of participant observation are valid? Evans (1988) reminds us of the very important point that any method is, to a degree, valid when the knowledge which it constructs is considered by stakeholders to be an adequate interpretation of the social phenomena which it seeks to understand and explain.

Are there disadvantages to reliance on observation? One danger, perhaps, is privileging face-to-face interactions over less localised relations that remain beyond the view of the researcher in the field (Gupta and Ferguson 1997). To believe only what we see would be to make the serious mistake of denying the existence of structures such as social class, or communicative processes such as internet relationships, that occur 'off-stage'.

Whether we seek to count, to gather complementary information, or to understand the context of place more deeply, the key to taking observation seriously is being attentive to detail as well as acknowledging our positions as researchers. Those recognitions imply that we are aware of both our place within the social relations we are attempting to study and the reasons we have the research agendas that we do (White and Jackson 1995).

KEY TERMS

complete observation
complete participation
controlled observation
'culturally safe'
fieldnotes
observation
observer-as-participant
panopticon
participant-as-observer
participant observation
positioning
primary observation
secondary observation
uncontrolled observation

REVIEW QUESTIONS

1. What are some of the ethical considerations that arise from observation?
2. In what ways might access to a specific social setting (for example, a netball club, an industrial workplace) be achieved for research purposes?
3. In what ways does observation involve more than seeing?
4. Suggest some ways in which our presence might influence interactions in the research setting.
5. Why, and in what circumstances, might one opt for participant observation as a research method?

SUGGESTED READING

A helpful guide for students using participation in dissertation research is provided by Ian Cook's 1997 chapter in *Methods in Human Geography: A Guide for Students Doing a Research Project* (Edited by R. Flowerdew and D. Martin, Longman Publishers). Another accessible resource, with concepts interleaved with research examples, is Mel Evans' chapter in Eyles and Smith's 1988 book *Qualitative Methods in Human Geography* (Polity Press).

7

Reading Texts and Writing Geography

Dean Forbes

CONTENTS

CHAPTER OVERVIEW

The aim of this chapter is to discuss different approaches to reading texts as a basis for writing geography. 'Reading texts' is an important tool of geographic research. Most studies make some use of critical readings of textual material, and some rely extensively on the analysis of texts. The chapter first discusses the content analysis approach to reading texts, before moving on to reading landscapes as if they were texts. The focus then shifts to contemporary approaches to reading texts through consideration of deconstruction and semiotics. Recent applications of deconstruction and semiotics in geography are illustrated by case studies of critical readings of magazines, Aboriginal landscapes, film and urban symbols. Finally the chapter summarises some of the recurring methodological characteristics of these approaches, and critically comments on the strengths and vulnerabilities of these methodologies for writing geography.

WHAT ARE TEXTS AND TEXTUAL ANALYSIS?

Eclecticism in theory or methodology has never been as sinful in geography as it is in some academic disciplines. Thus contemporary geographers bring a wide range of approaches to their research. Textual analysis is one kind of approach that geographers have found can be usefully applied to their own subject matter (Barnes and Duncan 1992). It draws upon research traditions from the humanities, which are qualitative in nature. Of course, the reading of printed texts has long been an essential part of geographic investigation. **Texts** contain factual information, theoretical interpretations, methodologies, and much more. Printed texts such as books, manuscripts and maps have been the primary means by which scholars in literate societies record and store knowledge. But in recent years some important changes have occurred to the way in which texts are conceptualised and used.

First, it is now recognised that 'texts' are much more diverse than simply books and manuscripts, and that significant interpretation of the world is contained in the analysis of diverse forms of text (for example, so-called **signs**) through such approaches as **semiotics**. Second, it is also increasingly recognised (though some would strongly dispute this), that reading texts is problematical. Geographers have always stressed the importance of a critical reading of texts. Some have embraced the sociology-of-knowledge approach, which stresses the context in which texts are written, and points to the importance of social and political events in shaping interpretation. However the emphasis is now on techniques of 'deconstructing' texts to reveal multiple meanings, ideologies and interpretations.

A common dictionary definition of a text is a form of printed matter such as a book or manuscript. That is not the meaning of text used in this chapter. Instead, contemporary usage of text refers to both written matter as well as other combinations of images which have some form of cultural significance. Thus texts might include conventional forms of printed material, such as nonfiction, fiction or a blend of the two ('**factoid**' or '**faction**'). But texts can also include maps, photographs, paintings, film and multimedia images (for example, Bell [1992], Hargreaves [1991], Holland [1991], McCracken [1991], Pawson [1991] and Roche [1991]). Even oral traditions and music might be considered texts, an example being the Aboriginal songlines so eloquently described by Bruce Chatwin (1987).

APPROACHES TO READING TEXTS

Human geographers define as one of their missions the interpretation of the connections between humans and the built and natural environment. Regardless of geographers' intentions in pursuing their version of the mission of geography, nearly all human geographers engage in the construction of some form of **representation** of human society and the built environment. This representation is inevitably simplified and abstract. More controversially, it is also invariably partial, fragmented, subjective and capable of multiple interpretations (or readings).

Geographers are not alone in seeking to construct representations of humans and their environment. Travellers record observations in their diaries; journalists convert their observations of political incidents or natural disasters into written, oral or visual 'news' items; novelists seek to provide a context within which to set their stories; and screenplay writers fashion images of places to enhance the visual impact of their films. Regardless of whether these texts represent primary source material, such as administrative memoranda or personal diaries, or interpretative secondary material, such as the scribblings of academics, researchers find them a rich lode for mining. **Intertextuality**, or the construction of new texts by reference to other texts, is a deeply embedded cultural practice in western societies. In other words, it is common for authors of texts to take as their departure point the textual representations of other authors, emphasising just how dependent authors are on those who have come before them and the cultural context in which they operate.

In the following pages, three approaches to reading texts will be discussed: **content analysis**, reading landscapes, and semiotics and deconstruction.

Content analysis

Content analysis is one of the most basic techniques for examining a text. It involves determining the importance of certain features or characteristics in a text, and then carrying out a search for them in the text (Box 7.1). For example, content analysis is often employed to determine whether newspapers show bias in their coverage of electioneering, or it can be used to determine the ways in which people (for example, women), places (for example, Japan) or issues (for example, debate over a World Heritage listing) are represented, or imag(in)ed through television programs. Most analyses are quantitative: they count the number of words or the centimetres of newspaper column space devoted to a particular issue, or the number of times particular images appear in magazines, or the minutes of television time devoted to those kinds of images. Software packages such as *NUD*IST* provide a computerised mechanism for identifying some of these characteristics of written texts (these are described more fully in Chapter 8).

As an example of content analysis, a study by Myers, Klak and Koehl (1996) examined media coverage of the wars in Bosnia and Rwanda with the intention of exploring how and why the press represents Rwanda as 'other'. Over a four-year period to 1994 the American newspapers surveyed gave much more extensive coverage of the Bosnian conflict (14,114 articles) than the one in Rwanda (560 articles). Moreover, there were great differences in the language used to describe the conflicts: in Rwanda the emphasis was on tribal and gang savagery, while in Bosnia the conflict was portrayed as more civilised. The differences in the reporting of both conflicts reveal underlying prejudices about the nature and significance of human tragedy in different parts of the world.

Box 7.1: Approaches to content analysis

1. The texts chosen for content analysis need to be carefully selected and sufficiently comprehensive in order to create credible findings.
2. The categories of characteristics for observation/investigation need to be clearly defined and not overlapping.
3. Measurement in content analysis can be the number of appearances of a characteristic (for example, a word or phrase), the amount of time or space occupied, or the frequency or intensity of the observation.
4. Computer software such as *NUD*IST* can be employed to format the text, assign characteristics to it, and retrieve the text.

Source: based on Kellehear (1993, pp. 34–8) and Bouma 1996 (pp. 77–82).

Reading landscapes

The meaning of texts for the purposes of 'reading' has been stretched by the anthropologist Clifford Geertz (1973), who has argued that culture is a text and can be read as one might read written material. Geographers have been encouraged by this to extend their interests into reading the landscape as if it were a text (Duncan 1987; Duncan and Duncan 1988; Dunn 1995). Though revived by association with the new discipline known as cultural studies, geographers familiar with earlier forms of cultural geography were also urged to read landscapes as texts. Pierce Lewis (1979) codified this into seven axioms (see Box 7.2). Embedded in these axioms is an assumption that there is a single culture which can be identified and described through the reading of the landscape. Current researchers would find Lewis' axioms overly simplistic.

Box 7.2: Seven axioms for reading the landscape

1. Culture is unintentionally reflected in the ordinary **vernacular** landscape.
2. All objects in a landscape reflect culture, and most are of approximately equal importance.
3. Ordinary landscapes are often overlooked in favour of significant symbolic structures.
4. History is vital to understanding contemporary landscapes.
5. Cultural landscapes need to be understood in their geographic context.
6. The biophysical environment is essential to reading the cultural landscape.
7. Objects in a landscape convey messages, but these messages are not always clearly expressed.

Source: based on Lewis (1979).

The task of interpreting meaning in texts is known as **hermeneutics**. In contrast to the approach of Lewis, contemporary approaches to reading texts (or reading the landscape) argue that written texts and cultural productions, such as landscapes (few of which are unaffected by human activity), share characteristics such as embedded meanings; that texts and productions are used by their authors as symbols and to communicate messages; and that there is great variability in the way in which both are interpreted by 'readers'. Thus texts and landscapes can be approached, or read, in the same way. This general approach, therefore, assumes there will be more than one valid reading of the landscape.

Semiotics and deconstruction

Daniels and Cosgrove (1988) point out the long history of human efforts to interpret symbols and imagery, noting a revival in studies of the imagery embedded in art at the turn of the century. The identification and description of symbols became known as **iconography**, whereas **iconology** referred to the search for the meaning of such symbols. Contemporary hermeneutical approaches to reading texts have more in common with semiotics, the study of systems (or the language) of symbols and signs, than traditional cultural geography (Duncan 1987). As Mitchell (1986, p. 8) argues, images must be read

> as a kind of language; instead of providing a transparent window on the world, images are now regarded as the sort of sign that presents a deceptive appearance of naturalness and transparence concealing an opaque, distorting, arbitrary mechanism of representation, a process of ideological mystification.

In the terminology of semiotics, signs are composed of **signifiers** or images, whereas the meanings derived from these are referred to as the **signified**.

Deconstruction of texts and the analysis of **discourses** (the intellectual currents into which they fit) provide an important, but not exclusive, set of approaches to the analysis of texts. This form of analysis reveals that it is naïve to assume that texts have a single meaning which is intended by the author and is read and understood by all. Instead there are multiple layers of meaning that can be derived from a text. Texts invariably contain internal self contradictions which expose them to multiple interpretations, any or all of which might be quite unintended by the author. According to this logic, the reader of this chapter may derive an understanding of it at odds with that intended by the author, and probably quite different from other readers. However, not every aspect of every reading of a text is different: shared meanings and interpretations can be found within texts, creating what are termed 'textual communities' (Duncan and Duncan 1988).

Deconstruction places significant emphasis on the problematical aspects of language. That is, language is not a precise tool; when incorporated in a text it does not embody some unique meaning. This assumption that language can have a specific and unique meaning for all can be described as essentialist. Instead, deconstructionists believe meaning is a product of the collage of surface features of language, such as the interrelation between signifiers. Hence texts do not have a single, deep meaning which can be revealed by experts but be invisible to others. This argument goes much further than the sociology-of-knowledge perspective, which stressed the influence of the context of the author and reader on the meaning of a text. Deconstructionists query the very depth of meaning in the first place.

If there are legitimate multiple meanings to a reading of a text, it follows that deconstruction has a subversive intent, for this approach questions the authority with which anyone might claim to have determined the 'correct' interpretation of a text. In other words deconstructionists are critical of the hegemonic or dominant view, principally because it helps to maintain power structures. 'The function of any ideology in power is to represent the world positively unified. To challenge the regimes of representation that govern a society is to conceive of how a politics can transform reality rather than merely ideologize it' (Minh-Ha 1991, p. 2). For example deconstruction provides the approach for questioning who determines what is the most meaningful literature (the **canon**) within a discipline, or what account of a place is the most authoritative. Naked power is critical to this view of reading texts, as the primary meaning, it is often argued, is determined by those with authority or power.

Deconstructionists also point to particular characteristics of texts. It can be argued that what is absent from a text may be just as significant as what is present. For instance, a text, such as a map, which discusses the cultural landmarks of an Australian city but identifies nothing of significance to the Aboriginal residents must be understood through its absence of reference to its minority population. A critical insight of the deconstructionist approach is to highlight the naturalising function of texts and discourse. By ignoring some things, yet elevating and situating others, a sense of what is perceived as natural and normal is created. Drawing out absences, or the way in which representations are constructed, challenges these assumptions of the natural.

Overwhelmingly the theories and techniques of reading texts have blossomed in a milieu largely untroubled by sharp expressions of cultural and ethnic diversity. However the rise of postcolonial movements, which challenge the views emanating from the privileged and hitherto hegemonic centres of the West, have fused with deconstruction and other variants of reading texts to add an important orientation to this form of analysis. Postcolonial approaches place their emphasis on resistance and the expression of dissent from the totalising discourse of global economic development (or underdevelopment). **Postcolonialism** is an expression of the voices otherwise unheard, 'the recovery of the lost historical voices of the marginalised, the oppressed and the dominated' (Crush 1993, p. 335).

READING TEXTS: CASE STUDIES

Qualitative research methods need to be applied with a good deal of sensitivity to the particular circumstances of the research. Thus it is more constructive to explore 'reading texts' through case studies, and seek explication from them rather than through further abstract discussion. The techniques of critically reading texts are shamelessly eclectic, borrowing practices from both within the discipline of geography and the humanities and social sciences more broadly.

Reading the *National Geographic*

Deconstructing the written text is a means of reassessing what we think we know, of searching to understand why it is that we think about particular things in the way that we do. Some written publications are more important in shaping public opinion than others, being very widely circulated or having a high level of public credibility. Among geographical publications few, if any, have been as important an influence on people's views of the world as the *National Geographic*. Its written text, or perhaps more importantly the photographs and the captions to them, are imprinted upon many memories. My own recollections go back to early primary school days, where leafing through the *National Geographic* was *de rigeur* during weekly visits to the library. By the late 1980s, fifty million people were estimated to read each issue.

Several critical readings of *National Geographic* have sought to deconstruct images transmitted through the magazine (see Lutz and Collins 1993; Rothenberg 1993). Rothenberg's study focuses on *National Geographic* from its launch in 1888 through to World War II. This particular period was chosen because it was when the 'trademark photographs' of the magazine became its most memorable feature, a situation which persists today. Rothenberg's reading focuses on a number of themes embodied in the magazine, with a special emphasis on the photography.

Essentially the *National Geographic* saw its mission as the promotion of science and education. Thus, with an approach typical of its time, scientific information was gathered objectively. Nature and humans were conceptualised as separate. The role of 'man' was to explore and tame nature. This coincided with the imperial project, of which the *National Geographic* was a great supporter. Articles dwelt on the economic resources of the places featured, identifying climate, ports and the useful crops grown. The local population was assessed in terms of its contribution to production, such as rural work or adaptability to factories. By undertaking a scientific investigation of the economic resources of a region the magazine encouraged the exploitation of these resources to be seen as a logical or natural process.

National Geographic's depiction of women and general treatment of issues of gender and race were very controversial. Rothenberg (1993, p. 162) believes 'native' people were objectified by being considered part of nature. This promoted a form of racism, in which 'bronzed' women (though they were sometimes called merely brown) could be photographed naked, or at least topless. White women were never treated in the same way, always being photographed fully clothed. The body as art object was already well established in American aesthetic discourse, which allowed *National Geographic* greater latitude in photographing female nudity, with its strong erotic undertones, than other magazines of the time. This remained the case as long as its focus was on non-Western women and was associated with classic art.

The magazine romanticised the so-called 'wild woman'. This was based on a belief in the equivalence 'of primordial nature with wild woman, creating and sustaining analogies between the temptations of unknown, unexplored land and unknown, unexplored women' (Rothenberg 1993, p. 157). Yet an implicit hierarchy of races was evident in magazine treatment of different peoples: Japanese and Chinese were clearly regarded as superior to other non-Westerners. There were also notable absences from the *National Geographic*: black Americans were hardly ever photographed (Rothenberg 1993, pp. 113, 168), revealing the selectivity of images of particular places.

The methodology of Rothenberg's study was shaped by the theories employed, without explicit comment in the research paper on the techniques used. The choice of the *National Geographic* is justified because of its importance—evidenced by its remarkable success. The period of study is defined as from the magazine's inception up to World War II, though there is no indication of whether all issues of the magazine were 'read', or simply sampled in some way.

In contrast, the study by anthropologists Lutz and Collins (1993, p. xiii) reads the *National Geographic* since World War II, focusing on the photography, and critically deconstructing the image of the magazine as 'a simple and objective mirror of reality'. This is a book-length study of three parts, each of which is preceded by a summary of the methods employed. The book begins by examining the parent National Geographic Society through its declared philosophy and the way in which its photographic imagery is processed, in order to establish the context within which the magazine is produced. The main study—which constitutes the second part of the book—focuses on the photographic images themselves. Some 594 photographs were randomly selected for critical analysis, one per article, from articles on 'non-Western' countries published between 1950 and 1986 (Lutz and Collins 1993, pp. 87–9). These photographs formed the basis of the authors' textual readings. Finally a group of 55 white Americans were recruited to look at 20 recent photographs and asked to respond to them. The respondents were selected in an endeavour to achieve a gender balance and diversity of social experience. In addition, the sample was designed to include an older Vietnam War era cohort and a younger post-Vietnam cohort to distinguish the social impact of the Vietnam War on the magazine's images (Lutz and Collins 1993, pp. 223–27). The study explores the way in which *National Geographic*, while intending to portray non-Western cultures, also reveals much about the values of 1950s 'middle America'. An underlying tension is revealed between the desire of the magazine to inform its readers about different peoples and places, and the need to validate core American values associated with race, gender, progress and modernity.

Because of its extensive readership, the *National Geographic* merits a careful and critical reading. Its written articles, and especially its photographs, have been influential in shaping perceptions of the non-Western world. Next is an examination of the way in which the reading of multiple texts contributes to a fuller understanding of the construction of one particular place.

A postcolonial reading: Aboriginal people and the city

Jane Jacobs (1996) illustrates the postcolonial/colonial dialectic with an examination of constructions of Aboriginal peoples' tourist sites in the city of Brisbane. Her readings encompass four different texts (though she does not use that term): a Brisbane City Council brief for an ecotourism centre; a tourist site, Mount Coot-tha; an Aboriginal walking and art trail; and Aboriginal mappings of sites in Brisbane. The choice of four texts allows her to explore different constructions of Aboriginality as objects for the tourist gaze. Postcolonialism in the work of Jacobs (1996, p. 161) is 'not so much about being beyond colonialism as about attending to the social and political processes that struggle against and work to unsettle the architecture of domination established through imperialism'.

The reading of the published documents on the ecotourism centre reveals the way that the City Council intended constructing 'a marketable image of how the Aboriginal culture has coexisted with the natural environment for several thousand years' (Jacobs 1996, p. 132). The document is read to reveal internal contradictions. For example it stresses the authentic quality of the display. This is disputed by Jacobs (1996, p. 134) who considers it to have a 'highly contrived and excessively commodified' quality, a product of its commercial aims. She then shifts her attention to Mount Coot-tha. Jacobs' (1996, pp. 138–42) reading of a turn-of-the-century tourist guidebook is used to speculate about the way in which the city vista was imagined, Brisbane being situated in its colonial context as an imperial city transported to the southern hemisphere.

'Indigenous tourings' provide the motif for the second half of the study, reflecting Jacobs' concern with postcolonial themes. In this case her interest is in the way in which Aboriginal peoples have constructed and Aboriginalised Brisbane's landscape. The first example of this is the formation of a tourist space in the form of a walking trail on the slopes of Mount Coot-tha. New artworks, such as rock paintings, are distributed along the trail. Jacobs (1996, p. 144) reflects on the hybrid iconography of the trail which mixes 'traditional significations of Aboriginal land knowledge as well as signifiers of colonial territorialisations' (or the territory taken over by the colonial authorities). In contrast to the planned ecotourism centre, the trail was planned and built by Aborigines and thus 'it has created place out of space' (Jacobs 1996, p. 145). On a grander scale the 'art trail is one of a number of recent place-making events in Australia which have unsettled, if not usurped, the surety of colonial power' (Jacobs 1996, p. 146).

Finally, Jacobs turns to maps as a source of insight. Maps, of course, are a form of text, at one time more central to geography than they are now, but nevertheless significant forms of representation (Pickles 1992). 'Deconstructing the map', the title of a paper by Harley (1992, p. 232), shows 'how cartography also belongs to the terrain of the social world in which it is produced. Maps are ineluctably a cultural system'. Several studies have drawn attention to the map as an instrument of colonial power; there is irony, therefore, in their postcolonial use by Aborigines. As Huggan (1995, p. 407) has noted, while 'the map is ironized ... as a visual analogue for the inflexibility of colonial attitudes and for the "synchronic **essentialism**" of colonial discourse, it is celebrated ... as an agent of cultural transformation and as a medium for the imaginative revisioning of cultural history'.

Jacobs' examination of Aboriginal mapmaking is an example of post-colonial reconstruction (a follow-on to deconstruction) which parallels the building of the Aboriginal walking trail. One map of the trail is located on a signpost at the start of the pathway, and another version is etched into rock. The same artist has produced a map of Brisbane which is intended 'not only as the logo for this site but as a city-wide signifier for places and sites with Aboriginal content and meaning' (Jacobs 1996, p. 151). The map of the trail is interpreted as containing a hybridity, or a doubleness, 'which destabilises the boundaries between Self and Other, colonial and traditional, authentic and inauthentic' (Jacobs 1996, p. 151). Clearly, the maps have a very powerful significance for Jacobs, challenging the certainties of colonial discourse.

To sum up, Jacobs' methodology has several distinctive features (see also Box 7.3 on p. 139). It is based on the examination of a variety of textual sources, each of a different kind, which are supplemented by interviews with some of the key individuals involved. However the methodology used in the study is not specified, de-emphasising the importance of empirical investigation. Only fragments of the method can be inferred from the text, in which it is partially revealed. As Duncan (1987, p. 473) stated, 'urban semiologists .... feel no such attachment to empiricism, asserting the primacy of the human intellect to create explanatory systems'. Instead, explicit theory becomes almost a substitute for a transparent, empirical methodology, framing Jacobs' study—in this example—within a powerful emphasis on a postcolonial reading of text.

Representations in film

The cinema is a significant means by which geographies are represented. Los Angeles freeways or the hilly inner streets of San Francisco are familiar because they often feature on television screens or in the cinema. Satellite television coverage of spectacular news stories or major sports events disseminate highly processed and edited images of the places affected. Yet relatively few scholars have looked at film as a source of insight into geography, or at film's impact on place and space, or in other words the dialectical relation between the two.

Film is of profound importance in the modern world. The techniques of film-making, such as the ability to vary space-time relationships in framing scenes, means it has an immediate impact on the senses (its so-called **haptical quality**) whilst transcending realism. The tendency for film's influence to leak into the landscape, mixing cinematic imagery into everyday events, means that the imagery of the cinema or television infiltrates community perspectives on society and space (Clarke 1997, p. 3). Approaches to reading the cinema (as text), according to Clarke (1997, p. 7), are triangulated by semiotics, psychoanalysis and historical materialism. In other words, the body of critical theory which informs the reading of cinema shares common roots with the other forms of critical analysis discussed in this chapter.

Take the case of the cinematic city, or the way in which the city is depicted in contemporary cinema. In a recent volume devoted to cinema and the city (Clarke 1997, pp. 10–11) two main themes predominate: whether the city is portrayed as utopian or dystopian (as in film noir, where the underbelly of the city is highlighted); and the relationships between the cinematic and extra-cinematic city.

Cinematic representations of the city can offer informed insights of a kind hard to find in any other form. A concern with postcolonial representations of the Vietnamese city led me to the films of Tran Anh Hung, a Vietnamese-born, Paris-based director and writer (Forbes 1999). Hung wrote and directed two recent feature films, *The Scent of Green Papaya* (1993) and *Cyclo* (1995). Both were set in Saigon/Ho Chi Minh City, but three decades apart. *The Scent* follows a young servant who moves from the country to the city and goes to work in a middle-class household in the early 1950s, eventually falling in love with the head of a second household in which she later finds work. *Cyclo*, by contrast, is the violent story of a young trishaw rider (or cyclo) and his sister, whose poverty forces them into a criminal gang, he as a thug and she as a prostitute.

Set in presocialist Saigon, in 1951 and 1961, *The Scent of Green Papaya* essentially represents Saigon in a utopian fashion. The camera seldom strays outside the household; fighting between the Vietnamese and French occurs off-screen, barely intruding on the dynamics of household life. While the servant's tasks are onerous, she is treated reasonably well, reinforcing the idea that the transition from country to city can be relatively painless. The relationship which develops with her second employer, in which she displaces his fiancée, implies that class and income barriers can be transcended by love.

In *Cyclo*, however, postsocialist Saigon has been renamed Ho Chi Minh City, and the dystopian vision dominates. The rampant market economy of post-economic-reform Vietnam has produced a vulnerable underclass, so that when the cyclo's vehicle is stolen, he is forced to become a criminal in order to pay his debts to his Boss Lady. The gang's involvement in the drug trade and the gratuitous violence which characterises their behaviour creates an image of a bleak, dangerous city out of the control of the authorities.

The cinematic city in the films of Tran Anh Hung resonates with the city experienced by the visitor, but with some important differences. Neither of Hung's portrayals jar; Saigon/Ho Chi Minh City is large and complex, so both the utopian and dystopian visions can be accommodated, being an essential part of the city's complexity. However, absent from his representations are any significant references to the outside influences on the evolution of the city. His postcolonial representation turns on the internal struggles within the family, the household and the community, whereas the visitor is much more likely to be struck by the symbols from the West, whether it be the struggle with the colonial French in his first film, or the corporate commercial invasion which marks the present city.

Though film can be read using similar techniques as textual analysis, there is much that is unique to film, not least because of its **mimetic** representation of character and space. The importance of film as a medium of geographic representation is unarguably of growing significance and therefore likely to assume a greater importance in contemporary geography. The importance of the distinctive styles of representation conveyed by films is also a useful means of exploring postcolonial expressions and constructions of place and identity.

Urban symbols

All cities are crammed with features rich in symbolic significance (Watson 1991; Winchester 1992). In the case of Canberra, the capital city of Australia, the symbolic landscape of the city centre has been consciously and painstakingly planned. The genesis of the structure of Canberra was a design competition from which was chosen a winning entry, laden with the symbolic landmarks thought appropriate for a youthful settler nation with grand aspirations (one entry to the competition envisaged a duplication of the palace of Versailles). On top of this was grafted layer upon layer of rethought images, creating a contemporary city intended to have a significant place in the Australian imagination (Forbes 1993; Freeland 1995; Game 1988). How could we then go about reading Canberra's landscape as a text in order to understand the city? How has the symbolic landscape of Canberra been constructed? What are the links between prominent contemporary conceptions of nation and identity? How inclusive is this symbolic landscape? Are there any concessions to postcolonial sensitivities and any symbolism of a multicultural society?

Canberra has at least two public identities. One is as the home of the Commonwealth Government and the institutions which support it. Thus when journalists talk of Canberra doing this or that they are using a **metaphor** for government behaviour. At the same time, Canberra is a small city of 299 000 people (1996); yet Canberra as a community is far less frequently in the public eye. Because of the symbolism of the city's landscape, and its association with Australia as a nation, reading it provides an illustration of an approach to urban semiotics.

Much, but by no means all, of the major symbolic edifices in Canberra are clustered in the area known as the Parliamentary Triangle. Here are located some of the most significant formal institutions in Australia: the Parliament, High Court, and the National Gallery. The structure of the triangle had its origins in a metaphor of an amphitheatre. The stage is occupied by the Parliament, on Capital Hill, the highest point within the triangle. Descending downwards to Lake Burley Griffin one passes the major institutions of the state, the slopes constituting the auditorium, and the galleries represented by Black Mountain and Pleasant Hill.

The metaphor has a natural logic, reflecting the ideological centrepieces of a democratic society, such as the parliament. Yet it invites a critical reading. How powerful is the institution of parliament given the strength of the executive in the Westminster system, or the influence of the very wealthy in a capitalist economy? And what of the people, who presumably inhabit the auditorium? The green spaces of the Parliamentary Triangle do not throng with people, they tend to be rather deserted, except for occasional cyclists or joggers. It is ironical that the auditorium should be so uninviting to the people.

The Parliamentary Triangle contains many other symbols of significance. These include the War Memorial, and Anzac Parade, along which are many smaller memorials to battles. The Parade draws a line between Parliament and the midpoint of the base of the triangle. This positioning brings the military to the foreground. It has an ideological function, naturalising the link between the military, the institutions of democracy, and the cluster of symbols of Australian identity. Yet a critical reading of these symbols would question this version of the significance of military history to national security, for few of these wars posed any direct threat to Australia. Nor, might it be argued, should the connection between war and national identity be constructed in the way implied in the symbolism of the city. Australian identity is a more complex entity than is represented in the Parliamentary Triangle.

A critical reading of the Parliamentary Triangle would investigate the absences from the symbolic landscape. Of significance is the lack of symbols associated with the multicultural (or multiethnic) population of Australia and—perhaps more striking—the virtual absence of recognition of the Aboriginal peoples of Australia. The forecourt of Parliament House is a mosaic of Aboriginal design, but this is the sole major symbol within the Triangle. Even then, the artist, Michael Nelson Tjakamarra, later took away the central stone, thus removing the symbolic status of the design and leaving it without meaning (Jacobs 1996, p. 146). Equally, symbols associated with women are largely absent from this landscape, which often draws comments from visitors about its phallocentricity. Thus the absences from the landscape speak of exclusivity, not inclusiveness, and Australian identity focused on a narrow range of symbols.

Because of the prominent role that Canberra has within Australia, it is particularly important to distinguish between the various kinds of reading of the city. The analysis above approaches a reading of Canberra from the perspective of urban semiotics (see also Game 1988). But

Canberra is also read by others, with different outcomes. Residents, for instance, generally see in it the quality of life afforded by a well-planned, modest-sized city.

The media, as I mentioned above, treat Canberra as synonymous with government. The Canberra community, it follows, is invariably portrayed as grey bureaucrats. The city is often contrasted with Sydney, the post-modern version of urban utopia. Cooke asks (1997, p. 53) 'Have you ever been to Canberra? It's a theme park called Boring World'. Sydney is rich in diversity and excitement, a real city; Canberra is 'an entire huge boring suburb of fully employed, house-bound Mormons with no taste in architecture'. Varying the metaphor to extract the maximum from the joke, 'Canberra is congealing custard on stale jam roly-poly'.

Representations of Canberra as a boring, small town, sadly contrasted with the glamorous metropolis of Sydney, echo representations of Melbourne as cultural centre contrasted with the parochial small-town-ness of Adelaide. This kind of reading of Canberra seeks to challenge its right to represent any symbols of contemporary Australia, implying that only Sydney has iconic status. It reproduces the ideology in which Australian identity is increasingly seen as synonymous with Sydney.

Methodologically speaking, the Canberra case study points to ways of reading (or **decoding**) the landscape. It highlights the importance of the identification and use of metaphor, the naturalising role of symbols in the landscape, and the importance of absences. Finally it draws attention to the complexity of who is reading the landscape, the context in which they are writing, and the hidden agendas in their work.

CONCEPTUALISING TEXTS

There is no single set of methodological guidelines for reading texts in geography. The theoretical approaches outlined at the beginning of the chapter, and the applications outlined in the case studies, borrow from a number of intellectual traditions. They are **pastiches** which appropriate from intellectual currents as diverse as French poststructuralism (deconstruction), studies of the grammar of signs (urban semiotics) and symbols (iconology), Geertz's approach to anthropology, and the empirical writings of the new cultural geographers. A consolidated set of recurring characteristics is presented in Box 7.3, simply to represent the overall flavour of these approaches.

Box 7.3: Reading the readings—some recurring methodological characteristics

1 Reading texts can be based on examination of a single text or, more commonly, of multiple, overlapping texts which help establish differences and similarities.

2 Texts are broadly defined, including government documents, academic publications, tourist guidebooks, maps, film, television, and the landscape; in fact any medium which can be interpreted for its meaning.

3 Most studies are highly theorised analyses with reference to a core group of structuralist and poststructuralist theorists (for example, Baudrillard, de Certeau, Deleuze, Derrida, Foucault, Kristeva).

4 Critical analysis uncovers alternative meanings from texts through the identification of self-contradictions, essentialist assumptions, absences, the existence of **multiple voices**, metaphors, taken-for-granted constructions (such as binary assumptions) and rhetorical devices.

5 Readings have subversive intentions as they seek to challenge and destabilise conventional interpretations, including appearances of naturalness, which repress, distort, mystify or oversimplify.

6 An emphasis on uncovering 'postcolonial' images exposes alternative representations to mainstream 'colonial' or 'imperial' images, through raising the question of who is entitled to authorship.

7 Research methodologies are generally unspecified, and only partly revealed by a reading of the interpretative text.

The eclecticism which characterises these approaches does not mean the methodologies used are invalid, or less than rigorous. The essence of postmodernist approaches is the exuberant synthesising, or juxtaposing, of diverse intellectual traditions. To put it more strongly, this reflects an implicit rejection of the notion of a single methodology for reading texts, or the idea that there is a single figure of authority who can provide a theoretical justification for this kind of analysis.

Moreover, eclectic approaches to reading texts do not require or result in an acceptance of **relativism**, in which any methodological approach is acceptable because there is no clear way of distinguishing the merits of one compared to another. Methodologies need to be well constructed and transparent if they are to produce trustworthy results (see Chapter 3). Essentially, most research based on readings is methodologically unsophisticated, the complexity and rigour of the work coming instead from the theoretical framework.

Nor should it be assumed that all research questions can be best investigated through a 'reading' approach. A strategy for choosing the methodology is necessary: is reading texts an appropriate approach to the particular question of concern, or is there a better way of going about the research?

The broadening appeal of deconstruction and other methods of textual analysis has been bitterly resisted by proponents of less obviously theoretical forms of textual criticism. As Mark Davis (1997, pp. 155–83) notes in his critique of cultural elites in Australia, the 'baby boomer generation' (itself a less than meaningful concept in this context) has been rather dismissive of new forms of literary and artistic criticism. For many, a critical reading of texts throws too much emphasis on theory and critique. It is not so much intertextuality which critics resent, rather the loss or diminished reliance on other practices. Novelists complain that the writing of novels is sacrificed at the expense of more critique, while geographers voice concern about the loss of the fieldwork tradition or the decline of participant observation.

It must be recognised that a methodology centred on reading texts has its vulnerabilities. Ethical issues must be dealt with. The 'reading' involves the critical review of the work of another person. There is a temptation to oversimplify and distort the views of an author in the researcher's enthusiasm to build a new argument. It may imply simply ignorance, or it may (mistakenly) imply an intention to deceive (as in willingness to be a vehicle of imperialism). It often ignores (perhaps intentionally) what is knowable at a particular point of time in a particular place. In other words, it privileges the researcher (i.e., the reader) over the previous author.

Authorship itself is a vulnerability, especially for those concerned with postcolonialism. It is implicit in scholarly research that the 'privileged academic' is able to undertake a more authoritative reading of a text than everyday readers. Moreover, questions of who is entitled to read texts and create or critique representations are often raised in these discourses. Is it appropriate that a scholar of Anglo-Celtic background seek to critique an Aboriginal representation of Aboriginal society? Should an Australian academic try to read an Asian landscape? These kinds of questions require reflection before embarking on research, but they also offer the bleak prospect of infinite contemplation and investigative paralysis.

Kevin Dunn (1995) is critical of the reading of the landscape approach. Text as landscape, he argues, is a metaphor. Though metaphors are pervasive in discourse, the text-landscape metaphor has some serious limitations. For example, unlike printed texts, landscapes are in a constant process of change. An emphasis on the signs embedded within the landscape ignores the functional role of the built environment. This echoes Jon Goss' (1988) critique that the emphasis on the symbol displaces concern about the significant material or administrative role of the city.

Duncan (1987, p. 481) also expresses concerns about the value of the inward-looking theory of abstract semioticians; the same end could easily be accomplished without the 'denotative' signifying systems which require an investigation of the 'pure grammar of spatial relations' (Duncan 1987, p. 475). He thus concludes that 'while the semiotic goal of decoding should be retained, much of the theoretical framework of the position should be dismantled'.

Nevertheless, it is important to acknowledge the insights which semiotics (or deconstruction) can bring to a discipline such as geography with its modern roots in logical positivism (discussed in chapter 9). These approaches drive home the point that texts are neither neutral (nor natural), that the texts and the way we read them are subjective and not objective, that readers determine different meanings from the text, not all of which are intended by the author, and that there are considerations of power and politics implicit in the readings of the texts. The aim of 'reading texts' is, therefore, to denaturalise them, and explore the multiplicity of meanings within them, and the power relationships which those meanings reinforce. We must also accept that these styles of reading texts provide no sense of closure. They do not result in a definitive statement, a nailing of the meaning of a text. Reading texts is a schematic for critical thinking. It is a journey with no end. It produces another generation of written texts which undermine (or at least challenge) the original, but which themselves can expect to be overturned by the next generation of textual readers.

KEY TERMS

canon
content analysis
decoding
deconstruction
discourse
eclecticism
essentialism
factoid
faction
haptical quality
hermeneutics
iconology

iconography
intertextuality
metaphor
mimetic
multiple voices
pastiche
phallocentricity
postcolonialism
relativism
representation
semiotics
sign
signified
signifier
texts
vernacular

REVIEW QUESTIONS

1 What kinds of research issues could be best pursued through a reading of a written text? Think of a specific example. How many texts would be analysed? What would be the purpose of such a study?
2 What kinds of problems would a deconstructionist identify in the use of a content analysis approach to texts?
3 Would a postcolonial approach to reading texts have some distinctive characteristics? What would they be?
4 How would an approach to the reading of film as a text in geography differ from a reading of a written text? It might be helpful to consider this question with reference to a book and a film linked in some way through their content, place or time (for example, compare the film 'Trainspotting' with Irvine Welsh's 1996 book of the same title, or consider a recent movie about US suburban life such as 'American Beauty' or 'Pleasantville' and compare it with a book/text dealing with similar material, such as Kunstler's 1994 book *The Geography of Nowhere*).
5 What are the limitations of an approach that uses semiotics or deconstruction to analyse a landscape?

SUGGESTED READING

Crush, J. 1993, 'Post-colonialism, de-colonization, and geography', in *Geography and Empire*, eds A. Godlewska and N. Smith, Blackwell, Oxford.

Duncan, J. 1987, 'Review of urban imagery: urban semiotics', *Urban Geography*, vol. 8, no. 5, pp. 473–83.

Harley, J.B. 1992, 'Deconstructing the map' in *Writing Worlds: Discourse, Text and Metaphor in the Representation of Landscape*, eds T. Barnes and J. Duncan, Routledge, London.

Jacobs, J.M. 1993, 'The city unbound: qualitative approaches to the city', *Urban Studies*, vol. 30, nos. 4–5, pp. 827–48.

Kellehear, A. 1993, *The Unobtrusive Researcher: A Guide to Methods*, Allen and Unwin, Sydney. See Chapter 3 especially.

Myers, G., Klak, T. and Koehl, T. 1996, 'The inscription of difference: news coverage of the conflicts in Rwanda and Bosnia', *Political Geography*, vol. 15, no. 1, pp. 21–46.

8

Computers, Qualitative Data and Geographic Research

Robin Peace

CONTENTS

CHAPTER OVERVIEW

This chapter aims to provide a brief introduction to the use of computers in qualitative human geography research. It focuses on a description of some of the basic aspects of computer assisted qualitative data analysis (CAQDA) systems and suggests two reasons for developing CAQDA skills as part of a researcher's tool kit. A brief summary of strengths, weaknesses and future directions for computer use in human geography research concludes the chapter.

INTRODUCTION

Sexy and chaotic

'Qualitative data are sexy ...' or so Miles and Huberman (1994, p. 1), well-known writers in the field of qualitative analysis in the social sciences, report. Researchers who use qualitative data have noted that such data are 'well grounded', comprise 'rich descriptions and explanations of processes', and are 'sexy' (Miles and Huberman 1994, p. 1). Well-grounded data draw from the experiences of informants. They 'tell it like it is' and provide the human detail of stories told about the contours of rich lives. Sexy data include the 'mysteries' and 'secrets'. Qualitative data can be construed as exciting, challenging and desirable. In this chapter, I ask whether or not computers can help us deal with these sexy data. Can computer software help in doing the analysis of 200 pages of interview transcripts, of video clips or policy documents or oral histories? While some researchers have noted the sexiness of qualitative data, other researchers have noted that qualitative data are 'unstructured, even chaotic' (Richards and Richards 1994, online), and are perceived to be 'airy-fairy'—not real research results. I also ask whether computer software can help contain potential chaos or manage the messiness of qualitative data and assist with data processing and organisation, and with managing tasks. The principal claim of this chapter is that whether you find your data sexy or chaotic, using computers can help.

FAQs—Frequently Asked Questions

There is a number of other, more mundane questions that novice researchers need to know about computers and analysing qualitative data, including such things as: How many kinds of software are there? Which one should I choose? How long does it take to learn how to use the software? Will the computer do my analysis for me and how will I know if it has got it right? These questions are personally motivated and you will not necessarily find all the answers you are looking for in this chapter. You will, however, find a very firm acknowledgment that **computers do not do analysis** —not for you, not for me, not for anyone. The most sophisticated software is merely one of many tools in a researcher's tool kit. You will also find at the end of this chapter a list of additional references and sources that will provide further guidance on where to find answers, assistance and help with making decisions.

Who else is doing it?

Computers are tools that are widely used by students and researchers in 'developed' countries. Qualitative research in disciplines as disparate as sociology, management studies, nursing and education have developed extensive literatures on qualitative methods that rely on computer assistance. Burgess (1996), Gahan and Hannibal (1997), and Fielding and Lee (1998), provide some recent examples of such work, as do the PhD medical research projects on chronic illness produced by Ayres (1998) and Buston (1997). Geography, on the other hand, although it has a well-developed reputation for computational work involving statistics and geographic information systems, is lagging behind in the uptake of CAQDA systems. Among the few pieces of geography research cited in the literature on CAQDA are Baxter (1998) and two other recent articles (Crang et al. 1997; Hinchliffe et al. 1997). For a critique of epistemological issues in computer use in geography, you might find it useful to refer to Peace (1998). Despite the small amount of published material, however, there is a growing awareness in geography that computer software has an important role in qualitative work.

TOOL KIT TECHNOLOGY

Computer software generally provides tools to help with the automation of four key tasks—storing, organising, retrieving and displaying symbolic information. More and more secretarial-type tasks are now expected of researchers—the typing, formatting, editing and sometimes even final production of documents, all of which have somehow become a part of the job of 'doing research'. To do these tasks quickly and efficiently you need good tools. Computers that can highlight errors in spelling and syntax and allow you to create tables and layout text, provide such tools. Increasingly, you will be expected not only to type and edit but also to manage and manipulate your own electronic database in quite sophisticated ways. You therefore need your own kit of specialised tools and specialist knowledge. You need hands-on familiarity with a computer system and the ability to navigate hardware and software. At the very least you need basic word-processing skills.

Before the 1990s (but since the late 1960s) computing in geography was largely associated with **GIS** (Geographic Information Systems), with *SPSS* (*Statistical Programs for the Social Sciences*)™, and *SAS* (*Statistical Analysis System*)™ software (Earickson and Harlin 1994; Griffith, Desloges and Amrhein 1990; Shaw and Wheeler 1994; Thompson 1992). These are software systems with the capacity to deal with the statistical calculations and manipulations of numeric or digitised 'real world' data—migration statistics, demographic data, census information or satellite readings. In GIS systems such as *ARCVIEW*™, the software facilitates links between graphic files (digitised pictures of some kind, such as maps) and **attribute databases** (such as information on the location or movement of some factor) in order to identify patterns or changes over time. *SPSS* and *SAS* are specialised statistical programs that are used to perform complex calculations and convert data from numeric to graphical display forms.

More recently, human geographers, along with other social scientists, have looked for computer support for analysing unstructured, qualitative data. Such support has been forthcoming in the form of specialised databases that have been 'custom built' for social science qualitative research. You can get access to complex and useful computer support for qualitative data analysis through so-called CAQDA systems.

What are CAQDA (Computer Assisted Qualitative Data Analysis) systems?

CAQDA is a generic title for a range of software systems that are specifically designed to handle unstructured, qualitative data. They are different from the data handling systems that have been familiar to geographers in the past. Over a decade ago, Forer and Chalmers (1987, p. 37) identified nine 'types of [geographical] activity' that could be effectively handled by computers. These included statistical analysis, modelling, word processing, graphics, cartography, image processing, visual imagery, remote sensing, and (artificial intelligence) concept handling. In the 1990s the range of activities has been extended to include word searching, data coding, data storage and retrieval, memoing, graphic mapping, hierarchical tree building, concept building, and reflexive report writing (see the Glossary at the end of this volume for explanations of some of these terms if they are unfamiliar to you).

Computer software that is designed to assist researchers involves a three-way relationship between the 'researcher', the research 'process' and the 'hardware and software'. The individual skills, attributes and desires of the researcher inform the kind of research that is undertaken, govern the researcher's actions and influence the kind of analysis that is performed. These two elements (i.e., 'researcher' and 'research processes') are in turn influenced by technology. The researcher uses technology to support or enhance the processes in which s/he is engaged. CAQDA systems, no less than other software, rely on this three-way relationship. Electronic technology can now facilitate a wide range of qualitative research processes. **Word searching**, a relatively simple task that most word processors can perform with fairly rudimentary instruction from the researcher (and therefore a relatively low level of skill) is at one end of the range. **Concept building**, which is a more sophisticated task that requires specialised software and considerable researcher training, exists at the other. The following two examples (see Boxes 8.1 and 8.2) illustrate some of the differences between using computers for basic CAQDA functions such as 'word searching' and using specialist CAQDA software. In each example characteristics of a hypothetical researcher and research scenario are outlined as well as the useful technology, the skills needed by the researcher and the advantages of computer use.

Box 8.1: Some characteristics of research based on CAQDA basic word searching

CAQDA: example 1—research using 'word search' capacities	
The researcher	May be new to research, perhaps working on an Honours or Master's thesis.
The research process	Involves qualitative analysis of transcribed in-depth interview data derived from a relatively small group of respondents.
The researcher's question	For example to find out whether or not the interview respondents refer to a particular term, such as 'homelessness', and the frequency of references.
Technology needed	
Hardware	Basic computer (desktop or portable Mac or PC).
Software	Word-processing program (for example, Microsoft Word™, Word Perfect™, Claris™ or Lotus™).
The researcher needs	Access to a computer, ability to use a keyboard and mouse (or willingness to learn), ability to use computer tools such as 'search' or 'find' function keys.

Ways in which the research process is enhanced through use of the computer:

- Transcribed text can be entered electronically and quickly as a 'document'.
- The 'find' or 'search' tool in the word processor can be used to look for key words.
- The 'search' returns a complete record of all occurrences.
- Large amounts of text may be searched very quickly.
- Material can be electronically transferred to another document without re-typing.

Box 8.2: Some characteristics of research using concept building by means of qualitative data analysis software

CAQDA: example 2—concept building using specialist software	
The researcher	Is likely to be experienced in qualitative research, possibly working at postgraduate level.
The research process	Involves qualitative analysis of transcribed in-depth interviews with a larger group of respondents (say up to 200 or more) but may also involve intensive specialised analysis of a small focus group.
The researcher's question	For example to investigate the importance of place for young people with persistent ill health (see Buston 1997).
Technology needed	
Hardware:	Computer with 486 capacity (desktop or portable Mac or PC).
Software:	QDA program (for example, *QSR NUD*IST™*, *ATLAS/ti™*, *the Ethnograph™*, *the Text Collector™*).
The researcher's needs	Access to an appropriate computer, software, institutional and user support, ability to learn new skills, and an ability to understand the function and limitations of the software.

Ways in which the research process is enhanced through use of the computer:

- Transcribed text can be imported and exported electronically from word-processing program to CAQDA program and back again.
- The 'find' or 'search' tools in the CAQDA program are very sophisticated and can be used to identify a range of relationships and ways in which key words or significant data occur.
- Categories and indexing systems for codes and texts allow for comprehensive retrieval.
- The CAQDA program allows for additional data to be associated with the text, such as memos, notations, diagrams of key relationships, **hierarchical trees** and graphic relationships.
- Some CAQDA programs allow for interfacing with quantitative databases such as *SPSS*.
- Very large amounts of text may be searched very quickly.
- Data can be effectively coded and records of all the codes kept and used for **data retrieval**.
- Concepts can be developed out of the coding processes because the coding can be flexible and fluid through the early stages of the analysis.
- Data can be multiply coded so that ideas and concepts can be examined from a number of different angles.
- The ability to retain records of coding patterns and coded material makes it possible for the researcher to rethink and recode—to work recursively.
- Subcoding patterns can be developed from major coding axes so that the detail of concepts can be examined.
- Theories can be built up, tested, dismantled, rebuilt and re-examined while the software retains records of the versions, trials and hunches.

Computer technology can be used to support a range of tasks that a qualitative researcher might wish to perform, but why would you choose to? In the next section we look at this question.

Reasons for using computers for qualitative research

I suggest there are two broad reasons for choosing CAQDA systems. First, using the computer allows you to automate a number of clerical tasks and to work more confidently with larger blocks of data. You might think about these two aspects as 'fast-tasking' and 'upscaling'. Second, they encourage versatility in approaches to research and encourage you to work more interactively with data—to use the software to support **theory building** and testing.

Fast-tasking and upscaling

In order to get fast results from word processing you need to be able to use the keyboard reasonably efficiently, and to have the confidence to find your way around the icons, menus and instructions of the word-processing program. Then you can shift data without retyping, correct spelling and syntax errors, search for keywords, edit, and print. If you wish to use a purpose-built program, such as *NUD*IST*, *ATLAS/ti*, the *Ethnograph*, or the *Text Collector*, there is a specific learning curve associated with each program which can not be circumvented. Once learnt, however, you can engage in 'multisite, multimethod studies .... that may combine qualitative and quantitative inquiry ....[and be] carried out by a research team working with comparable data collection and analysis methods' (Miles and Huberman 1994, p. 2). It is claimed that good CAQDA systems allow you to 'collect, compare and organise varied documents; code and explore them, discover patterns; search for words, phrases in text; explore cases, record emerging ideas and theories; relate theories to data, link with statistical records; report results, [and] provide evidence' (*QDA Resources* 1998, online).

The speed with which examples of text can be retrieved is a significant time saver for the researcher, but so is the ability to manage unwieldy databases. Some software enables you to keep tabs on data that is off-line, that is not directly 'machine-readable' and therefore cannot be handled directly by the software. Off-line data might include bibliographies, untranscribed interview data on audio-tape, movie or video clips, static images or material on compact disk (CD). Being able to make memos and annotations relating to off-line data can successfully extend your database without costly scanning or transcribing processes being involved.

Richards (1997, p. 429) cautions against opting for large-scale projects under the illusion that good research involves collecting vast amounts of data. She suggests that although the computer has this capacity for prodigious memory work, it should not be used lightly. CAQDA systems are as effective working with small databases as they are with larger ones. It is also clear, however, that the core assets of computers for qualitative research work are their capacity for organizational memory, and for **data storage**, retrieval and display. Being able to systematically file material, which can be retrieved on the entry of a keyword, puts computers in a different league from systems such as card indexes. CAQDA software that helps automate clerical tasks can help you work quickly and systematically (see Tesch 1989) and develop versatile and interactive approaches to your data.

Versatility and interactivity

Computer software can make data coding and categorising more flexible. Not only can you experiment freely with various categories, you can also code data more fluidly in different ways. You do not have to make up one coding system for your research project and then allocate data to one set of either/or categories. Programs such as *QSR NUD*IST* allow you to put the same segments of data under a number of different headings or categories and retrieve them through a range of indexing and searching tools that are activated by the codes the segments of data have been given. You can do multiple coding and then instigate complex inquiries of those coding patterns. This versatility can also enhance 'interactivity' between the researcher and the data.

Research, at its best, is a recursive process—one in which researchers interact intimately with their data. As Michael Agar (1986, in Crang 1997a, p.188) suggests, the process of coding and subdividing codes can be 'maddeningly recursive' because the categories that seem stable and appropriate one minute tend to break down as the research progresses, so that new categories must be invented. This recursivity, this bending back on itself, is one of the great and exciting strengths of qualitative research. The ability of much CAQDA software to facilitate detailed record keeping of these recursive processes is what enables software developers to claim that their product enhances theory building. A particular strength of several of the current software programs (especially *NUD*IST* and *ATLAS/ti*) is that they also enable relationships to be articulated graphically—either through their own 'tree' structures or through graphical interface with more recently developed software such as *Inspiration*™ and *Decision Explorer*™. Many people who think visually find it very helpful to be able to build their conceptual thinking graphically.

Apart from these broad arguments in favour of using computers for qualitative research, there is a vocational argument. Being able to do 'fast' research, being able to demonstrate a capacity to analyse more than a dozen in-depth interviews, and having a sense of the versatility and interactive skills involved in qualitative research also makes for a very marketable researcher. Having a wide range of skills as a software user enhances your qualifications. You should consider this seriously when making a decision about whether to invest time in learning to use a particular software program or not. If you are serious about considering the possibility of using computer software to help with qualitative analysis, it is imperative that you read widely and make an informed decision. A list of useful references is provided at the end of this chapter. World Wide Web (WWW) sites that give access to more technical, software-specific information are listed under 'Key Internet resources'. The following section gives a brief introduction to five generic types of qualitative software.

Different types of software

There is no recipe for choosing the software that will be best for your project. You need to bear in mind the kind of computer you are using, your budget, what access you have to software in your institution, your current skill level and the amount of time you have to learn new skills, and which kind of program is likely to be most useful to you. Rather than give any product specifications or recommendations I have summarised some of the key information in Box 8.3. Weitzman and Miles (1995) provide a comprehensive study of 24 different software programs (see also Fielding 1994 and Fielding and Lee 1998). Product homepages and mailing lists are usually the best sources for the most up-to-date information (see Key Internet resources).

Box 8.3: Different kinds of CAQDA software

- **Text retrievers**
  These kinds of software are great for doing text analysis where you need to hunt words, strings of words or characters and retrieve and batch them up into categories and groups. They include software such as *Metamorph*™, *Orbis*™, *Sonar Professional*™, *The Text Collector*™, *Word Cruncher*™ and *ZyINDEX*™.
- **Text retrievers/Textbase managers**
  These kinds of software perform similar functions to the above but have the additional capacity to facilitate the systematic organisation of words or characters into records based on fields and subsets. This software includes *askSam*™, *Folio VIEWS*™, *Max*™ and *Tabletop*™.
- **Code and retrieve**
  These softwares take the retrieval and management aspects a step further by incorporating capacity for attaching coding. Key words or codes can be applied to data chunks so that relevant sections of texts (rather than individual words or characters) can be retrieved according to a coding formula. Programs in this range include *HyperQual2*™, *Kwalitan*™, *Martin*™, *QUALPRO*™, and *The Ethnograph*™.
- **Code based theory building**
  Code based theory builders are designed with capacity for retrieval, coding, annotating, memo making, and cross-questioning. They usually have some capacity for graphic representation of coding structures and patterns and increasingly are compatible with quantitative software programs such as *SPSS*, so that you can use multi-method approaches. Included in this category are programs such as *AQUAD*™, *ATLAS/ti*™, *HyperRE-SEARCH*™, *QSR NUD*IST*™ and *QCA*™.
- **Graphic display**
  Graphic display and **conceptual mapping** software programs have been developed to interface with other CAQDA software programs so that coding patterns can be exported from one program and displayed in another. *NUD*IST* coding, for example, can be exported to *Inspiration*™. Other graphical display programs include *Decision Explorer*™, *MECA*™, *Metadesign*™, and *SemNet*™.

In choosing which of these software programs to use the decision needs to always be driven by the question: what can this software help me to do more quickly and more reliably than I could do with pen and paper? In the next section I briefly address some broader issues—the epistemological strengths and weaknesses of CAQDA systems—as well as summarising some of the more mundane frustrations and constraints.

STRENGTHS AND WEAKNESSES OF CAQDA SYSTEMS

Strengths

I have already suggested that using computers in qualitative data analysis can revolutionise the way you work by taking you away from the marking-up/cut-and-paste styles of analytic work that prevailed in pre-computer days. The most significant advantage is that you can integrate data entry much more closely with analytic processes. You can stay 'on-line' to do your analysis and have greater control over the proliferating paper work. The computer has a much more efficient 'memory' for filing and automates much of the drudgery of replicating information. Sophisticated software programs also allow for exploration, innovation and theorising. The time investment required to learn new software skills usually pays off.

Weaknesses

Computer technology is not universally accessible. Neither is the ability to link up to the internet, nor the ability to load up software to facilitate research. Computer technology requires a power supply (with all the concomitant assumptions of 'developed', industrialised economies), a supply of appropriate hardware and software, and a method of gaining access to whatever system support is offered as part of the program. It also requires a decision to commit to a project that entails capital costs, and it requires that all of these elements be in place before the research gets under way. Word processing programs are usually availbale for use in most research institutions, but not all provide or support more specialised software programs. Trying to finance the purchase of software and software training and support out of your own budget can be very costly. Trying to learn to use new software on your own can be very demoralising.

Generally I would suggest that there are four provisos to bear in mind if you are thinking of using computers for your qualitative research. First, you must have time to learn whatever system you propose to use. Sometimes that may take weeks—perhaps even several months. Second, you must have access to a computer system that can work with the chosen software. Third, you need some aptitude for interfacing with computers. Even for 'computerphobes' that aptitude can be cultivated by careful and patient teaching. Fourth, you have to be motivated to learn particular skills. For many undergraduates, specialist software programs may not be the most appropriate solution to data analysis problems. If you feel compromised by any of these provisos—if your time is short, if you do not have easy access to appropriate computer hardware, if you do not feel immediate rapport with computer technology—then it may be best to learn the basics now and come back to the specialist software later. Advantages and limitations of CAQDA systems are debated in recent literature such as Barry (1998), Bazely (1997), Crang et al. (1997), Fielding and Lee (1998), Hinchliffe et al. (1997), Kelle (1997a, 1997b) as well as in more established texts (see the readings suggested at the end of this chapter).

Undergraduates and graduates, as well as established researchers, need to remember that there are complex configurations of practical/technical and theoretical/political issues to bear in mind when assessing the role of computing in their own explorations of qualitative research. Computer software does offer exciting new developments and capacities, and it also offers computer users the chance to acquire marketable skills and a sense of conceptual confidence and flexibility that may be harder to achieve with other methods. Qualitative research using computer software demands, above all else, a high degree of researcher reflexivity. It is to be hoped that any full-scale move to using computers to make sense of qualitative research will entail on-going reflection and critique.

CONCLUSION—FUTURE DIRECTIONS

I hope that more geographers who experiment with CAQDA software will publish accounts of their experiences and methods and the field will be opened up for widespread discussion and debate. An exciting new development might see the production of purpose-built software for interfacing between GIS programs and CAQDA systems in much the same way that *NUD*IST* and *SPSS* programs are now compatible. Increasingly there will be a trend towards working across qualitative/quantitative boundaries to produce multi-method work that draws research insights from the widest possible background. Such work will depend on the capacities of computers to manage databases—to store, organise, retrieve and display symbolic information—in ways which make it possible to deal with data more efficiently and with greater versatility, to reduce drudgery and increase flexibility and to provide researchers with interactive, portable and valued skills.

KEY TERMS

attribute database
code and retrieve software
computer assisted qualitative data analysis software [CAQDAS]
concept building
conceptual mapping
data retrieval
data storage
fast-tasking
geographic information system [GIS]
hierarchical trees
listserve
text retriever software
theory building software
upscaling
word searching

REVIEW QUESTIONS

1 Qualitative data have been described on the one hand as 'rich', 'sexy', and 'well-grounded' and as 'chaotic' on the other. Discuss the ways in which these attributes are appropriate descriptors for qualitative data.
2 Discuss the ways in which qualitative and quantitative data are different and require different styles and technologies for their management.
3 Computers are frequently described as 'tools'. Discuss some of the specific characteristics of computers that are indicative of this.
4 The initials 'CA' in the acronym 'CAQDA' refer to the particular relationship that computers have to the analytic processes involved in research. The idea that computers 'assist' the analyst is a far cry from the notion that computers 'do research'. Discuss the relationships between data analysis and computer technologies in qualitative research.

5 Qualitative research is often described as a 'recursive' process. What is meant by recursivity? How do recursive approaches assist qualitative analysis? What might be an example of a recursive process in qualitative research?
6 Choose any one of the five generic kinds of CAQDA software (see Box 8.3) and use the Internet to explore the capacities of the software in that category. Compare your findings with someone who chose a different category. Discuss which kinds of software would be appropriate for any research project you know about.

KEY INTERNET RESOURCES

This internet resource list is subdivided into seven categories:

- Finding out more about CAQDA
- Finding out more about qualitative research
- Finding out more about computers
- Product pages for specific software packages
- Electronic journals
- Email discussion lists
- Internet sites for geographers

Finding out more about CAQDA

CAQDAS *(Computer Assisted Qualitative Data Analysis Software) Networking Project*, <http://www.soc.surrey.ac.uk/caqdas/> (3 November 1999).

Qualitative Data Analysis Tools: CTI Psychology Website, <http://www.york.ac.uk/inst/ctipsych/dir/qualitative.html > (3 November 1999).

Sociological Research Online, <http://www.socresonline.org.uk/socresonline/> (3 November 1999).

Using Computers in Sociological Research: Discussion Forum, *Sociological Research Online*, <http://www.socresonline.org.uk/socresonline/threads/computers/computers.html> (3 November 1999).

Finding out more about qualitative research

Association for Qualitative Research, <http://www.latrobe.edu.au/www/aqr/about/aboutaqr.htm> (3 November 1999).

Qualitative Analysis: what is it?, <http://info.ippt.gov.pl/~zkulpa/quaphys/QAnalys.html> (3 November 1999).

Qualitative Comparative Analysis, <http://www.nwu.edu/IPR/publications/qca.html> (3 November 1999).

Qualitative Research Resources on the Internet,
<http://www.nova.edu/ssss/QR/qualres.html> (3 November 1999).
QualPage! resources for qualitative researchers,
<http://www.ualberta.ca/~jrnorris/qual.html> (3 November 1999).
Text analysis software sources,
<http://www.intext.de/TEXTANAE.HTM> (3 November 1999).

Finding out more about computers

Exploring the Internet, Nicky Ferguson, *Social Research Update*,
<http://www.soc.surrey.ac.uk/sru/SRU4.html> (3 November 1999).
Glossary of Internet Terms, Enzer Matisse 1994-98,
<http://www.matisse.net/files/glossary.html> (3 November 1999).
PCWebopaedia, <http://www.pcwebopedia.com/index.html> (3 November 1999).

Product pages for specific software packages

Atlas/ti, <http://www.atlasti.de/index.html > (3 November 1999).
AQUAD—Analysis of Qualitative Data, <http://www.uni-tuebingen.de/uni/sei/a-ppsy/aquad/aquad.htm> (3 November 1999).
AskSam, <http://www.asksam.com/> (3 November 1999).
Code-A-Text, <http://www.scolari.co.uk/codeatext/codeatext.html> (3 November 1999).
Decision Explorer, <http://www.banxia.com/demain.html> (3 November 1999).
The Ethnograph v5.0, <http://www.QualisResearch.com/> (3 November 1999).
FolioViews, <http://www.folio.com/folio/Factsv41.cfm> (3 November 1999).
HyperRESEARCH 1.65 Software for Qualitative Data Analysis,
<http://www.researchware.com/hr165.htm> (3 November 1999).
Inspiration Software, Inc., <http://www.inspiration.com/> (3 November 1999).
KWALITAN, <http://www.kun.nl/methoden/kwalitan/kw-des-e.htm>
(3 November 1999).
METAMORPH Intelligent Text Retrieval,
<http://www.thunderstone.com/jump/Metamorph.html> (3 November 1999).
QSR Forum, <http://www.qsr.com.au/Training/QSRForum.htm> (3 November 1999).
Symantec (anti virus encyclopoedia),
<http://www.symantec.com:80/avcenter/vinfodb.html> (3 November 1999).
SAS Institute, <http://www.sas.com/> (3 November 1999).
SPSS Statistical Product and Service Solutions, < http://www.spss.com/>
(3 November 1999).
winMAX 97, <http://www.winmax.de/produkte.htm> (3 November 1999).
WordCruncher Internet Technologies, <http://www.wordcruncher.com/>
(3 November 1999).
Zylab International ZyINDEX, <http://www.provantage.com/PR_11043.HTM>
(3 November 1999).

Electronic journals

The following journals provide online access to articles dealing with issues of computer use in qualitative research.

The Qualitative Report: an online journal dedicated to qualitative research and critical inquiry, <http://www.nova.edu/ssss/QR/index.html> (3 November 1999).

Social Research Update, <http://www.soc.surrey.ac.uk/sru/Sru.html> (3 November 1999).

Sociological Research Online, <http://www.socresonline.org.uk/socresonline/> (3 November 1999).

ESRC Data Archive Bulletin, University of Essex, <http://dawww.essex.ac.uk/about/bulletin.html/> (3 November 1999).

Email discussion lists

Qual-software (email discussion group), <http://www.mailbase.ac.uk/lists/qual-software/> (3 November 1999).

QSR Forum (email discussion group), <http://www.qsr.com.au/Training/QSRForum.htm> (3 November 1999).

SQUERM—Supporting Qualitative and Ethnographic Research Methods, (email discussion list), <http://www.mailbase.ac.uk/lists/squerm/> (3 November 1999).

Internet sites for geographers

Internet Resources for Geographers, <http://www.utexas.edu/depts/grg/virtdept/resources/contents.htm> (3 November 1999).

Geography Resources: Software and Commercial Vendors, <http://www.utexas.edu/depts/grg/virtdept/resources/vendors/vendors.htm> (3 November 1999).

SUGGESTED READING

Barry, C. 1998, 'Choosing Qualitative Data Analysis Software: Atlas/ti and Nudist Compared', *Sociological Research Online*, vol. 3, no. 3, <http://www.socresonline.org.uk/3/3/4.html> (3 November 1999).

Coffey, A. and Atkinson, P. 1996, *Making Sense of Qualitative Data, Complementary Research Strategies*, Sage, Thousand Oaks.

Denzin, N. and Lincoln, Y. eds, 1994, *Handbook of Qualitative Analysis*, Sage, Thousand Oaks.

Dey, I. 1993, *Qualitative Data Analysis: A User-Friendly Guide for Social Scientists*, Routledge.

Fielding, N. and Lee, R. 1998, *Computer Analysis and Qualitative Research*, Sage, Thousand Oaks.

Gahan, C. and Hannibal, H. 1998, *Doing Qualitative Research Using QSR NUD*IST*, Sage, Thousand Oaks.

Kelle, U. 1995, ed, *Computer-Aided Qualitative Data Analysis: Theory, Methods and Practice*, Sage, Thousand Oaks.

Miles, M. and Huberman, M. 1994, *Qualitative Data Analysis: A Sourcebook of New Methods*, Sage, Thousand Oaks.

Weitzman, E. and Miles, M. 1995, *Computer Programs for Qualitative Data Analysis: Software Sourcebook*, Sage, Thousand Oaks.

9

Writing In, Speaking Out: Communicating Qualitative Research Findings

Lawrence Berg and Juliana Mansvelt

CONTENTS

CHAPTER OVERVIEW

In this chapter we examine the process of 'writing-up the results' of qualitative research in human geography. Our aim, however, is to contest the simplistic understanding of the relationship between research, writing, and the production of knowledge which arises from describing the process in this manner. Indeed, the very phrase 'writing-up' implies that we are somehow able to unproblematically reproduce the simple truth(s) of our research in our writing. In this framework for understanding research, writing becomes a mirror that serves to innocently reflect the reality of research 'findings'. In contrast, we draw upon poststructuralist approaches to argue that writing is not merely a mechanical process that reflects the 'reality' of qualitative research findings, but, instead, writing constitutes in part how and what we know about our research. Writing is thus not so much a process of writing-UP, as one of writing-IN, a perspective which has significant implications for how research is conceptualised.[1]

STYLES OF PRESENTATION

Part of our argument in this chapter revolves around the idea that a number of powerful **dichotomies**—such as subject/object, researcher/researched, data/conclusions, and research/writing—currently exist and that these dichotomies structure our understanding of research. These dichotomies have developed historically as part of a positivist framework that we suggest, limits our understanding of the writing process. We thus feel it is important to present readers with a very brief history of the development of positivist thought in human geography before moving on to discuss an alternative approach to writing human geography.

Positivist and neo-positivist approaches: universal objectivity

The philosophical approach to scientific knowledge known as **positivism** was founded by Auguste Comte (1798–1857), a French philosopher and sociologist (see Gregory 1978; Kolakowski 1972). Drawing on the work of empiricists, Comte argued that scientific knowledge of the world arises from observation *only*. Positivists had a relatively singular conception of truth, and they had significant difficulties with issues and

1 Readers seeking specific 'how to' advice on stylistic conventions associated with the presentation of research are advised to consult Hay (1996), Stanton (1996), Hay, Bochner and Dungey (1997) or Kneale (1999) in conjunction with the conceptual material of this chapter.

questions arising from the relationship between truth and phenomena such as religion, ethics, morals and metaphysics.

The logical positivists, whose work developed as a critique of strict positivism during the 1920s, differed from the Comtean positivists in that their conception of scientific knowledge accepted the validity of more than just empirically verifiable statements. In this sense, Comte allowed only synthetic statements arising from empirically verifiable knowledge (that is, information available to the senses) as the basis of factual knowledge. The logical positivists, on the other hand, acknowledged the existence of two kinds of factual statements: *analytical* statements, the truth of which is contained in their internal logic (e.g., if A=B and B=C, then A=C); and *synthetic* statements, whose truth is proven through recourse to empirical observation (e.g., if we observe 500 cases of a specific chemical reaction between sodium and ammonia, then our empirical observations will lead us to conclude that this particular chemical reaction is a natural characteristic of mixing sodium and ammonia).

The logical positivists were criticised by Karl Popper (1959) who developed an alternative form of neo-positivist thought he termed 'critical rationalism'. Unlike the logical positivists, who focused upon verification as the basis of knowledge, Popper argued that falsification should be the basis for making decisions about the veracity of factual statements. His argument was based on the notion that we can never know for sure whether a particular hypothesis was true or not, but we do have the ability to ascertain if it had been falsified. Under critical rationalism, then, we never prove something but instead we can either disprove something (or prove it to be false through the process of falsification), or accept it to be contingently 'true' (until it has been proved false). According to Popper, scientific knowledge was much more contingent than conceptualised by the logical positivists, since present 'truths' were always open to future falsification.

Although few geographers have ever fully taken up the ideas of any single school of positivist thought, positivism of one sort or another has played a foundational role in the way that many geographers have come to understand the world. This became most explicit during the so-called 'quantitative revolution' of the 1960s (Gregory 1978; Guelke 1978), which saw geography develop as a broadly positivist 'science'. Geographers maintained a strict distinction between facts and values (Gregory

1978); they privileged (i.e., gave emphasis to) observational statements over theoretical ones (Berg 1994a); they made a strict distinction between **objective** and **subjective** knowledge; and they tended to universalise their findings across all contexts (Barnes 1989).

In adopting a broadly positivist model for their work, geographers also developed a specific approach to writing their research. They developed commonly used approaches or forms (or what we call **'tropes'**) in their academic studies that attempted to erase the authorial self from their written work. Similarly, they tried to create, through their writing, distance between themselves as researcher/author and their research objects. These tropes are most evident in the practice of writing in the third person—a practice that is still prevalent in much academic writing today. Most geography undergraduate students, for example, are still required to use the formal third-person narrative form in their essay assignments (for example, see School of Global Studies 1998, p. 7).

The implicit rationale for writing in the third-person—and for the subsequent erasure of the author and distancing of the author from the research object—is the need to maintain *objectivity*. Indeed, ideas of objectivity (as distanced observer) and impartiality form the keystone of the positivist model. In this model, the third-person narrative is not only acceptable—because knowledge is conceptualised as universal—it is also *required* in order to ensure impartiality and objectivity. It does this, ostensibly, by preventing the intrusion of authorial bias or subjectivity into the narrative in the rhetorical form of the first person. In discourse analysis and social semiotics, this form of writing is seen as an attempt to construct an 'objective modality' (Fairclough 1992). 'Modality' indicates the writer's *degree* of agreement with a particular statement. The use of objective modality, as opposed to the more explicit 'subjective modality', has the effect of removing the text producer from their text, while at the same time implying the author's full agreement with a statement. Objective modality thus connotes a facticity (i.e., a factual character) about the statement that is beyond question. This manoeuvre transforms interpretation into fact.

Writing in the objective mode is often accompanied by **nominalisation**, a process which further removes the writer from the text. Nominalization involves the transformation of adjectives and verbs into nouns (called derived nominals) (Fowler 1991, p. 79). An example of nominalization common to academic writing is the transformation of 'research' (as a verb) into 'research' (as a noun). What was once a process involving participants (researchers and research 'subjects') becomes a thing, lacking active agents who are involved in a set of social relations. Another form of nominalisation involves the use of nouns that designate actions and processes. For example although they are nouns, terms such as 'a call', 'the inquiry', or 'the campaign' imply or designate some form of action or process. Nominalisation offers much ideological purchase because of the way it *reduces* information. As Fowler (1991, p. 80) suggests, this process is easily illustrated if we compare the nominalisation 'allegations' 'with the fully spelt-out proposition "X has alleged against Y that Y did A and that Y did B [etc.]"'. Nominalisation deletes information about the participants (the agent and affected participant), the modality (the writer's degree of agreement with the statement), and the circumstances. Nominalisation thus has two key outcomes: *mystification*, because it permits concealment, hiding the participants, indicators of time or place, and modality; and *reification*, whereby processes assume the status of things (Fowler 1991, p. 80). Ironically, while positivist epistemology is founded on the idea that any statement can be contested through recourse to empirical evidence and logic, the language used to communicate research findings within this framework implies an unquestionable accuracy.

Objectivity is constituted in opposition to subjectivity, and it is founded on interrelated, mutually constitutive and highly gendered notions of rationality, disembodied reason, and universality (Berg 1994a, 1997; Bondi 1997). In this regard, scientific geographers, as Trevor Barnes and Derek Gregory (1997, p. 15) suggest, imagined themselves

> as a person—significantly, almost always a man—who had been elevated above the rest of the population, and who occupied a position from which he could survey the world with a detachment and clarity that was denied to those closer to the ground (whose vision was supposed to be necessarily limited by their involvement in the mundane tasks of ordinary life).

From this disembodied vantage point, objective knowledge looks the same from any perspective—it is monolithic, universal and totalising. This concept of un-located and disembodied rational knowledge draws on powerful metaphors of mobility (the researcher can move to any and all perspectives) and transcendence (the researcher is not part of the social relations she or he is examining, but instead can rise above them to see everything) for its rhetorical power to convince readers (see Barnes and Duncan 1997; Haraway 1991).

Interestingly, bearers of this approach have shown ambivalence with regard to their own writing practices. On the one hand, they have explicitly acted as if language has no impact on meaning; on the other hand, they have implicitly acknowledged—through their insistence on writing in the third person—the significant role that language plays in constructing knowledge and meaning.

Post-positivist approaches: situated knowledges

Although the positivist-oriented science model dominated during the quantitative theoretical period of the 1960s and 1970s, its **hegemony** (or dominance) is now contested in geography by post-positivist approaches (for example, Berg 1994b; Barnes 1993; Dixon and Jones 1996). In particular, feminist **poststructuralist** writers have been keen to confront the universalism, mastery and disembodiment inherent in positivist notions of objectivity. Such ideas, they suggest, play a significant role in marginalising those who do not fit into dominant conceptions of social life.

Over the last decade feminist writers in particular have adopted various spatial metaphors—'politics of location' (Anzaldúa 1987; Frankenberg and Mani 1993), 'cartographies of struggle' (Mohanty 1991), 'power-geometries' (Massey 1993), and '**situated knowledges**' (Haraway 1991)—in order to criticise masculinist and Eurocentric concepts of universal knowledge (also see Berg and Kearns 1998; England 1994). As Liz Bondi (1997, p. 248) points out:

> they contribute to critiques that challenge the mastery of dominant knowledge systems, that emphasize the corporeality of knowers, reject the radical split between mind and body associated with the notion of transcendence and that argue that all claims to a singular, universal position are fraudulent. They argue instead, partly via their distinctive spatial metaphors, for critical, situated knowledges.

Donna Haraway's (1991) evocative metaphor of *situated knowledges* provides perhaps the most useful trope to contest universalist forms of knowledge. She argues that within dominant ideologies of scientific knowledge, objectivity must be seen as a 'God Trick' of seeing everything from nowhere. She proposes a different concept of 'objectivity', one that attempts to situate knowledge by making the knower accountable to their *position*. All knowledge is the product of specific embodied knowers, located in particular places and spaces: 'there is no independent position from which one can freely and fully observe the world in all its complex particulars' (Barnes and Gregory 1997, p. 20). Research that draws upon situated knowledges is thus based on a much different notion of objectivity from that posed by positivists, and this conception of objectivity also requires a different form of writing practice.

Recent work by Isabel Dyck (1997) provides an excellent example of the importance of 'position' and the kind of 'truths' situated knowledge might produce (also see England 1994). Dyck reflects on the importance of her own position as a white middle-class Canadian academic and the impact this had on two different research projects she did into the time-space strategies adopted by Indo-Canadian immigrant women in Vancouver. She found that immigrant women were more willing to speak with her about certain aspects of their lives than others because she was not seen as a threat to their own social networks and relationships within the Indo-Canadian community in Vancouver. At the same time, as an outsider, she was occasionally excluded from aspects of Indo-Canadian women's lives that were defined as culturally sensitive. Her research thus points to the specificity of position and the importance of recognising the politics of position in research processes (see Chapters 2 and 3 for additional material on these matters).

Poststructuralists and feminists also contest those approaches to inquiry that conceptualise writing and language as simple reflections of 'reality'. Instead, drawing upon the work of people like Ferdinand de Saussure (1983, originally published in 1916 after his death) and Martin Heidegger (1996, first published in German in 1927), they argue that 'language lies at the heart of all knowledge' (Dear 1988, p. 266). It should be made clear, however, that such arguments do not conflate reality and language—that is, they do not assume that language and ideas are the same as phenomena, objects and material things. Instead, arguments about the centrality of language express the fact that all processes, objects and things are understood by humans through the medium of language. Thus, while we might experience the very material process of hitting our 'funny bone' on a table in ways that do not necessarily entail language (for example, as pain), we come to understand the process and the objects involved through language (with categories such as table, funny bone, pain, etc.). Accordingly, language must be seen as not merely reflective, but instead as *constitutive* of social life (for example, Barnes and Duncan 1992; Berg 1993; Bondi 1997; Dear 1988).

The poststructuralist critiques of both the mimetic concept of language and of disembodied concepts of universal objectivity have significant implications for writing practices. If language is constitutive of knowledge and meaning, then it would seem to matter *how* we write our knowledges of the world. Likewise, if we are to *locate* our knowledge, then we must locate ourselves as researchers and writers within our own writing. Accordingly, poststructuralist writers reject the ostensibly 'objective' modality of writing their work in the third person. Instead, they opt for locating their knowledge—defining objectivity as something to be found not through distance, impartiality and universality, but through contextuality, partiality and **positionality**. However, as Gillian Rose (1997) has recently argued, given the difficulty in completely understanding the 'self', it may be virtually impossible for authors to fully locate themselves in their research. Notwithstanding such difficulties, it is possible for authors to go some way towards locating themselves within their work. Certainly the first step is to reject the third-person narrative, replacing it with a first-person narration of our essays.[2] Thus rather than writing ourselves out of our research, we write ourselves back in. This is, perhaps, one of the most important distinctions to be made between what we will refer to as the 'writing-UP' (distanced, universal, and impartial) and the 'writing-IN' (located, partial, and situated knowledge) models. Another significant difference arises from the ways that post-positivists conceptualise the relationship between observation and theory.

BALANCING DESCRIPTION AND INTERPRETATION—OBSERVATION AND THEORY

The role of 'theory' and the constitution of 'truth'

We argue in this section that there currently exist a number of powerful dichotomies—observation/theory, subject/object, researcher/researched,

2 It is appropriate to note however that it can be difficult institutionally to present some forms of research this way. For instance, social or environmental impact statements prepared by government departments or consulting firms will frequently not list authors, where indeed the 'we' would be poorly defined. Moreover, neither group would want individuals to be sued for their opinions.

data/conclusions and research/writing—that structure our understanding of research. It is important to remember that these dichotomies are not recent developments in Western thought. Instead, they arose within the long history of dualistic thinking in Western philosophy (Berg 1994a; Bordo 1986; Derrida 1981; Foucault 1977; Jay 1981; Le Doeff 1987; Lloyd 1984; Nietzsche 1969). Indeed, right from the time of the early Greek philosophers, dichotomies have structured Western understandings of the world (Cavendish 1964). In the sixth century BC, for example, the Pythagoreans constructed their 'table of opposites', which posited the concepts of enlightenment, goodness, orderliness and determinacy in opposition to those of darkness, badness, disorderliness and imprecision (Cavendish 1964; Lloyd 1984). These dichotomies became racialised and gendered through a long historical process of developing a singular Eurocentric and masculine concept of rational thought. For example through a process that Susan Bordo (1986) terms the 'cartesian masculinisation of thought', Descartes' mind-body distinction came to define appropriate forms of knowledge. The mind was conceptualised as rational and it came to be a property of European men. The body was seen as irrational (have you ever heard the phrase 'mind over matter'?) and was associated with everything that was not European or masculine: women, racial minorities, and sexual dissidents for example. In other words, the mind was unmarked but the body was a mark of difference. In this way, 'the mutual exclusion of *res extensa* and *res cogitans* made possible the conceptualization of complete intellectual transcendence of the body' (Bordo 1986, p. 450). Further, Descartes' dualistic philosophy of knowledge formed the basis for the dominant present-day conception of objectivity as impartial, distanced and disembodied knowledge. Accordingly, Cartesian dualistic thinking forms the foundation of positivist thought (Karl Popper, for example, was a well-known adherent of the mind-body dualism).

As we have already discussed, positivist-inspired geographers privilege observational statements over theoretical ones. They also make a rigid distinction between objective and subjective knowledge. As with the observation/theory binary, the so-called 'objective' is valued at the expense of the subjective. Such hierarchically valued dichotomies form parts of a whole series of other binary concepts—including (but not limited to) mind/body, masculine/feminine, rationality/irrationality, and research/writing—which are interlinked through complex signification chains (Derrida 1981; Jay 1981; Le Doeff 1987; Lloyd 1984). In this way, for example, observation is equated with objectivity, mind, masculinity, rationality and research. On the other hand, theory is constituted as *lacking*, and it is associated with all those other negatively valued concepts: the subjective, the body, the feminine, the irrational.

Conceptualising one side of the binary as a *lack* of the other leads to a devaluation of the subordinate term. Thus, in the case of those positivists who conceive of theory as a *lack* of empirical observation, the ways in which theory constitutes our understanding of empirical 'reality' (in addition to explaining it) are underestimated. Indeed, positivist constructions of factual knowledge as phenomena that are available to the senses (empirically observable), tend to efface the very theoretical nature of positivist thinking itself. As we have already illustrated, positivism has a history (and although we have not discussed it, positivism has a geography associated with Europe). It is an epistemology—a theory of knowledge—that has developed relatively recently and has come to dominate contemporary intellectual life in the West. Nonetheless, it is not the only theory of knowledge; rather, it is one of many competing theories of knowledge. However, because it is dominant, or hegemonic, it rarely has to account for its own epistemological frameworks. With this in mind, we argue for 'recognition that we cannot insert ourselves into the world free of theory, and neither can such theory ever be unaffected by our experiences in the world' (Berg 1994a, p. 256). Observations are thus *always already* theoretical, just as theory is always touched by our empirical experience. Recognition of this relationship has important consequences for the way we write-IN our work (and WORK-in our writing).

Writing and researching as mutually constitutive practices

Metaphors which allude to research as 'exploration' or 'discovery' are hard to avoid, so taken for granted are their meanings (how often have you heard lecturers speak of their research in terms of 'exploring', 'examining', 'discovering', and 'uncovering'?). This is particularly the case with the so-called 'writing-up' of the research. The term writing-up powerfully articulates the written aspect of the research process in a way which engenders it somehow less significant and/or less problematic than other aspects of the research process. Writing-up is usually seen as a phase which occurs at the end of a research program; indeed, many textbooks about conducting research (this one included) include the section on 'writing-up' at the end of the book (for example, Flowerdew and Martin 1997; Kidder, Judd and Smith 1986; and Kitchin and Tate 2000). Writing is also often seen as a neutral activity and the 'writing-up phase' may be discussed in such a way that it appears to be merely a matter of presenting the results and conclusions in an appropriate format. We argue here that in writing research the researcher is not so much presenting her or his findings as *re-presenting* the research through a particular medium. Re-presentation speaks of the mediated character of the process of writing research.

Rather than reflecting the outcome of a particular research endeavour, we believe the act of writing is a means by which the research is constituted—or given form—and that this process occurs throughout the research process. For example keeping in mind the constitutive character of language, we can see that any attempt to write research involves a process of selecting categories and language to describe complex phenomena and relationships. But we are getting at much more than that here. Research and writing are iterative processes, and writing helps shape the research as much as it reflects it. At the larger scale of disciplinary practices and epistemological conventions, as we have suggested above, the way we conceptualise the author (as distant and impartial or as involved and partial, for example) has significant implications for the ways that the very processes of research itself can be understood. For us, then, writing is not so much a matter of writing-UP as of writing-IN, a perspective which has considerable implications for how qualitative research is conceptualised and undertaken.

It is unhelpful to assume that the 'writing of research' is a phase that occurs entirely at the end of a research program. Writing is not devoid of the political, personal and moral issues which are a feature of undertaking research. Further, the separation between 'fieldwork' and writing is artificial (Denzin 1994). Whatever the qualitative research technique utilised, some means of recording the researcher's interpretations, impressions and analysis must be used, and though such accounts may be recorded on tape or video, the words with which they are constructed are an integral part of the research, not simply a result, recollection or recording of it. The research cannot be separated from the labels, terms, or categories used to describe it and interpret it, because it is through these that the research is made meaningful.

For example, after I (Juliana) had conducted several qualitative interviews with local authority Economic Development Officers for my PhD research, I realised that using the word 'traditional' as a label for a certain form of local economic initiative was problematic. This was because definitions of 'traditional economic initiatives' were contested by officers and because the term appeared to position local authorities who undertook this type of initiative as 'old-fashioned' or 'not progressive'. I learnt a valuable lesson about the power embodied in words I had simply utilised from the literature on local economic development and about the kind of assumptions embedded in my own research agenda. This understanding enabled me to construct my questions (and consequently my entire research project) in a different way.

I (Lawrence) have had similar experiences of re-orienting research agendas in a recent project I undertook with two colleagues (McClean, Berg and Roche 1997). This research helped define some of the historical geographies of Maori tribal land losses in the Porirua region of New Zealand—but it also involved a research 'partnership' with Ngati-Toa Rangatira, a Maori tribal organization whose lands formed the focus of the research. In this instance, because of our commitment to a research partnership with Ngati-Toa Rangatira, the 'writing-up' of the research was only the initial phase of an iterative process of negotiating the production of knowledge between the three researchers and the *iwi* (tribe). I am not suggesting here, however, that negotiated knowledge is intrinsically better than other forms of knowledge. Instead, what I want to point out is that the process of making explicit the act of negotiation helps to make the research accountable in ways that are appropriate given the specificity of positions and power relations. All research is caught up with power relations, and to deny this is to deny an important aspect of knowledge production (see Chapters 2 and 3). Taking this process seriously has enabled me to completely rethink the role of writing in my intellectual endeavours (see McClean, Berg and Roche 1997).

Thoughts, observations and interpretations which occur during the research thus become an important component of any research endeavour, not because they record events or ideas but because they are a signifier of them (in this sense, they act to 'define' complex constellations of ideas and thoughts about the research in more simplified categories of knowledge). Whether interpretations are noted by way of a personal diary, log, video, or audio recording they can provide insight into the researcher's own speaking position and how this is articulated, challenged and modified through the research journey.

Writing-IN is not a matter of 'telling'—it is about knowing. The process of writing constructs what we know about our research but it also speaks powerfully about who we are and where we speak from. As we have suggested in a previous section, the detached third-person writing style so common in academic journals and reports implies that the researcher is omnipotent—that they have a perspective which is all-seeing and all-knowing. However, what may appear to be the truth spoken from 'everywhere' is actually a partial perspective spoken from some*where* and by some*one*. Knowledge does not, according to a post-structuralist perspective, exist independently of the people who created it—knowledges are partial and geographically and temporally located. As the researcher writes and inscribes meaning in the qualitative text they are actually constructing a particular and partial story. This applies to neo-positivist modes of writing (as discussed above) too, but may be less obvious as the partiality of such texts is often hidden behind universal terminology and impersonal language. Richardson (1994) suggests that writing creates a particular view of not only what we are talking about, but also of ourselves. Power is connected with speaking position through text, so a qualitative researcher should consider not only the standpoint spoken from in constructing a research account, but also the implications of their interpretations for those who may have been involved in the research and on the structuring and power relations in everyday life.

Because the practice of writing is not neutral, the voices of qualitative researchers do not need to hide behind the detached 'scientific' modes of writing. Such modes of writing position the researcher as a disembodied observer of the truth, rather than a (re)presenter and creator of a particular and partial truth. The researcher is an instrument of the research and accordingly we suggest they should acknowledge their position in ways that demonstrate the connection between the processes of research and writing. **Reflexivity** is the term often used for writing self into the text. Kim England (1994, p. 82) defines this as 'self-critical sympathetic introspection and the self conscious analytical scrutiny of self as researcher'. A reflexive approach can make researchers more aware of asymmetrical (i.e., where a researcher has more social power and influence than their participants) or exploitative relationships, but it cannot remove them (England 1994, p. 86; also see Rose 1997 and Chapter 2 of this volume). One way in which reflexivity can be encouraged in the writing-IN of qualitative research is by the use of personal pronouns (for example, I, we, my, our).

Alison Jones (1992) suggests that in academic spheres the use of 'I' has become the enemy of truth; it represents the insertion of 'emotion' which replaces 'reason' thereby creating a work of 'fiction'. However, using 'I' can make explicit the politics associated with the personal voice and draw attention to assumptions embedded in research texts. Yet reflexivity is not ensured by the insertion of self into the text through the use of the personal pronoun (indeed, merely inserting the personal pronoun into a text can hide power relations as much as it might make them explicit); it is also concerned with constructing research texts in a way that gives consideration to the voices of those who may have participated in the research. Reflexivity is about writing critically, in a way which reflects the researcher's understanding of their position in time and place, their particular standpoint and the consequent partiality of their perspective. This understanding of the dialogic nature of research and writing (in the sense of a 'dialogue' between various aspects of the research process) enables qualitative researchers to acknowledge in a meaningful way how their assumptions, values and identities constitute the geographies they create. It also provides an opportunity to play and experiment with writing as a way of knowing and representing.

Richardson (1994) believes that to write 'mechanically' shuts down the creativity and sensibilities of the researcher. She encourages researchers to explore text and genre in the (re)presentation of qualitative research through a variety of media, including oral and visual. Richardson suggests researchers experiment with diverse forms of the written word (prose, poetry, play, autobiography) and write research pieces which are not conventionally linear in format and structure. We support Richardson's (1994) metaphorical construction of writing as '**staging** a text' and wish to encourage geographical researchers to consider how they are (re)presenting the research 'actors', creating the plot, action and dialogue of a research 'tale'; how they are constructing the stage, the setting of the 'research' play; and to whom (i.e., the audience) the 'production' is aimed.

Of course, most undergraduate geography students will be required to write their research within a given format, the essay or report (for detailed advice on the conventions associated with these forms of writing, see Hay [1996], Hay, Bochner and Dungey [1997] or Kitchin and Tate [2000]). Nevertheless, it may be possible to convince your lecturer to allow you to produce another form of geographic representation: a play, a video, a poem, a short story, a poster-board, to name a few options. Despite writing constraints and 'staging' conventions imposed by self and audience (such as for an 'academic' publication or a thesis) we want to argue that there is no single correct way to 'stage' a text. By exploring the varied ways in which the text can be staged, and how in such staging different stories may be emphasised and other voices may come to the fore, the researcher has the potential to create dynamic and interesting research pieces which engage and challenge both writer and reader.

Issues of validity and authenticity

The interpretative nature of qualitative research has given rise to a considerable amount of debate concerning how the **validity** and authenticity of qualitative research accounts might be assessed (see Baxter and Eyles 1997). The reflexive writing-IN of research experiences and assumptions is not a license for sloppy research or monographs based solely on personal opinion. Rigour, integrity and honesty in writing-IN are no less important in qualitative research than they are in quantitative research. Works by writers like Mike Davis (1990), Gillian Rose (1993) and Cole Harris (1997) provide telling examples of critical qualitative research and rigorous analysis of socio-spatial relations.

The truth and validity of knowledge arising from geographical and social science research has been the subject of discussion for a considerable period of time. Almost two decades ago, for example, Harrison and Livingston (1980) suggested that conduct of research is necessarily linked to what it is possible to know and how we can know it. Poststructuralist thinking has also challenged the assumption of a singular truth and the privileging of certain claims to knowledge. Associated with this is what has been termed the 'crisis of representation' (Marcus and Fisher

1986). That is, doubts have arisen over researchers' authority to speak for others in the conduct and communication of research (Alvermann, O'Brien and Dillon 1996). In recent years, an increased sensitivity to power and control on the part of some qualitative researchers has encouraged a rethinking of research design and implementation (Glesne and Peshkin 1992, p. 10), and it has also meant a growing concern over the textual appropriation (how researcher's appropriate participants' voices) of data in the writing of research accounts (Opie 1992). More recently, Baxter and Eyles (1997) have argued that geographers need to be more explicit about how rigour has been achieved throughout the research process. The process of writing-IN qualitative research requires, therefore, that writers explicitly state the criteria with which a reader may assess the 'trustworthiness' of a given piece of research. Addressing this issue is difficult because it involves qualitative writers grappling with the tensions between the complexity and richness of information which emerges from qualitative research and the need to produce some sort of 'standardised' evaluation criteria (Baxter and Eyles 1997).

The poststructuralist challenge to a singular notion of truth and a growing awareness of issues of representation do present difficulties for assessing the validity of qualitative research. Much of the debate surrounding the validity of qualitative writing rests upon how terms such as 'rigour', 'validity', 'reliability' and 'truthfulness' are defined. A related issue is whether these definitions, which have often been used as 'objective' measures of the quality of quantitative research, are applicable to qualitative endeavours (see Chapter 3). Poststructuralist thinking casts doubt on foundational arguments that seek to anchor a text's authority in terms such as reliability, validity and generalisability (Denzin 1994).

A debate by two geographers in 1991 and 1992 issues of *The Professional Geographer* highlights some of the issues surrounding reading and evaluating texts and the validity of qualitative approaches. The exchange between Erica Schoenberger (1991, 1992), and Linda McDowell (1992) is significant because it was written at a time when in-depth interviewing was not used extensively in industrial and economic geography. This debate is interesting not only because it centred upon contested definitions of validity but because it raised issues of meaning and interpretation, audience, and representation through the language which constituted these articles. It demonstrates powerfully the multiple reading of texts and the care needed in constructing qualitative interpretation. Schoenberger sought to argue for the legitimacy of qualitative forms of interviewing, but in doing so suggested how quantitative definitions of reliability (defined by Schoenberger as the stability of methods and findings) and validity (accuracy and truthfulness of findings) might be applied to qualitative research. McDowell disagreed with Schoenberger's concept of validity as interpretation which is verifiable and which corresponds to some external truth. McDowell argued that Schoenberger's definition was based on a continued adherence to a positivist-inspired quantitative research model, as demonstrated by the language through which her arguments were constructed. Schoenberger (1992) addressed McDowell's criticisms by referring to issues of intended audience (positivist geographers), power and interpretation, suggesting McDowell had misinterpreted her claims.

If concepts such as validity are contested, how then is it possible to construct rigorous research texts? Clifford Geertz (1973) has argued that good qualitative research comprises 'thick' description. Such descriptions take the reader to the centre of an experience, event or action, providing an in-depth study of the context and the reasons, intentions, understandings and motivations that surround that experience or occurrence. Though an understanding of 'the heart of the matter' is not independent of researcher interpretation and it will differ according to researcher and 'researched' (Denzin 1994), we believe it is possible to produce 'thick' descriptions of the world in which we live. While it may not be possible to assess the authenticity of such partial descriptions, the validity of the interpretations upon which they are constructed can be examined. Mike Davis (1990, pp. 253–57), for example, provides a particularly compelling description of the make-shift prisons and holding centres for the urban underclass who make up the more than 25000 prisoners in 'the carceral city' found in a three-mile radius of Los Angeles city hall (see Box 9.1). His descriptions, meticulously researched and referenced, are as much novel-like evocations of city life as they are academic descriptors.

Box 9.1: Writing Fortress L.A.

The demand for law enforcement *lebensraum* in the central city, however, will inevitably bring the police agencies into conflict with more than mere community groups. Already the plan to add two highrise towers, with 200–400 new beds, to County Jail on Bauchet Street downtown has raised the ire of planners and developers hoping to make nearby Union Station the center of a giant complex of skyscraper hotels and offices. If the jail expansion goes ahead, tourists and developers could end up ogling one another from opposed high-rises. One solution to the conflict between carceral and commercial redevelopment is to use architectural camouflage to finesse jail space into the skyscape. If buildings and homes are becoming more prison- or fortress-like in exterior appearance, then prisons ironically are becoming architecturally naturalized as aesthetic objects. Moreover, with the post-liberal shift of government expenditure from welfare to repression, carceral structures have become the new frontier of public architecture. As an office glut in most parts of the country reduces commissions for corporate highrises, celebrity architects are rushing to design jails, prisons, and police stations.

An extraordinary example, the flagship of an emerging genre, is Welton Becket Associates' new Metropolitan Detention Centre in Downtown Los Angeles, on the edge of the Civic Centre and the Hollywood Freeway. Although this ten-story Federal Bureau of Prison's facility is one of the most visible new structures in the city, few of the hundreds of thousands of commuters who pass it by every day have any inkling of its function as a holding and transfer center for what has been officially described as the 'managerial elite of narco-terrorism'. Here, 70% of federal incarcerations are related to the War on Drugs. This postmodern Bastille—the largest prison built in a major US urban center in generations—looks instead like a futuristic hotel or office block, with artistic charms (like the high-tech trellises on its bridge-balconies) comparable to any of Downtown's recent architecture. But its upscale ambience is more than mere façade. The interior of the prison is designed to implement a sophisticated program of psychological manipulation and control: barless windows, a pastel color plan, prison staff in preppy blazers, well-tended patio shrubbery, a hotel-type reception area, nine recreation areas with nautilus workout equipment, and so on. In contrast to the human inferno of the desperately overcrowded County Jail a few blocks away, the Becket structure superficially appears less a detention than a convention center for federal felons—a 'distinguished' addition to Downtown's continuum of security and design. But the psychic cost of so much attention to prison aesthetics is insidious. As one inmate whispered to me in the course of a tour, 'Can you imagine the mindfuck of being locked up in a Holiday Inn?' (Davis 1990, pp. 256–57).

Communicating qualitative research is thus about choices, for example over what to present, to whom and how (Strauss and Corban 1990, p. 247). Though such choices are not always conscious and not necessarily made in circumstances of our own choosing, researchers can attempt to approach transparency in research (that is, accountability to their perspective and position) through reflexively acknowledging and making explicit those choices which have influenced the creation, conduct, interpretation and writing-IN of the research. Such choices are likely to be guided by principles of ethics, truthfulness and rigour (see Chapters 2 and 3). Transparency may make researchers, and the audiences for whom they write, more aware of the constraints on interpretation, of the limitations imposed by the 'textual staging' and of the implications of the former for research participants. For example the choices surrounding the use of research participant's quotes in a written text comprise far more than a simple matter of how, where, how many and in what form participants' voices are to be included. The inclusion of quotations raises issues of representation, authority, appropriation and power (see Opie 1992 for a discussion on how textual appropriation may be reduced). Kay Anderson (1999, pp. 83–4), for example, provides a particularly compelling description of the juxtaposition of identities and spaces in Redfern, an inner-city suburb of Sydney closely identified with spaces of 'Aboriginality' (see Box 9.2). Anderson weaves together quotes from research participants, evidence gathered from archival research and rich theoretical writing to argue for understanding Redfern as a hybrid space of porous and fluid identities and spaces. Her work manages quite subtly to acknowledge issues of power in research and representation, while simultaneously providing us with a vivid description of life on the Block.

Box 9.2: Rethinking Redfern—writing qualitative research

On the Block [a small area of Redfern, an inner-city suburb of Sydney] itself, in 1994 the dominant language group was Banjalang, but a wide range of other place-based dialect groups were present, including Eora (Sydney region), Wiradjuri (Nowra), Kamlaroi (Dubbo and Moree), and many other groups from throughout New South Wales and Queensland. In addition, approximately two-thirds of the total number of rent-payers on the block were women, a significant minority of whom (among those interviewed) were married or partnered to Tongans, Fijians, Torres Strait Islanders, and members of other ethnicities. In most cases the men did not live with the women, who supported their children and funded the periodic visits of husbands or partners, friends and relatives. Sociability has always been both fractious and friendly. Some tenants saw the Block as 'home'; others perceived their place of birth as home; most considered they had multiple homes, including Redfern. In the words of a tenant known on the Block as a community elder and who has since refused to leave it: 'Redfern is an Aboriginal meeting place. People come from all over the country to get news of friends and family'. Now a member of the housing coalition which has been formed to fight the AHC [Aboriginal Housing Company], the same woman recently stated that Redfern has long been a place where children taken from their relatives encountered relatives or information that would lead to those relatives (Melbourne Age, 1 February 1997). Another woman had this to say in an interview for a recent documentary: 'People who come from interstate or wherever make to Sydney. Their first aim is the Block because they gotta know who's who and where's where and where to go from here' (ABC TV 1997).

The 'traffic of relating' on the Block opens up fresh ways of thinking about home and community in a context of more widely invoked images of 'flight', 'flow', 'crossings', 'travel' and transnational exchange in contemporary cultural geography. Models of mixing that work with the idea that cultures are porous and fluid are particularly apt in relation to this case study. There is, as I have suggested, no 'pure' culture at Redfern, no crisp boundaries of inside and outside, even for so stigmatised an area. This is not only a methodological issue. It is also an epistemological problem in that the boundaries of researchable communities are not secure and areas never exist as discrete entities (Anderson 1999, pp. 83–4).

Jamie Baxter and John Eyles (1997) suggest the criteria of credibility, transferability, dependability and confirmability are useful general principles for guiding an evaluation of the rigour (trustworthiness) of a piece of qualitative research. They see these categories as broadly equivalent to the concepts of validity, generalisability, reliability and objectivity that have been used to evaluate the quality of quantitative research endeavours. We believe it is important to keep in mind the constructedness of these concepts, to avoid using them as universal assessment criteria and to avoid engaging in comparative analysis of qualitative studies. Notwithstanding this, Baxter and Eyles' principles have much to commend them in that they may assist in evaluating the internal consistency and rigour of a piece of research. The application of these principles should encourage researchers to explore and make explicit their own research agendas and assumptions and to elaborate on how they believe their research text constitutes the 'truth' about a particular subject. While transparency is not in itself ultimately achievable, if the conscious reflexive writing produces qualitative texts which are open to scrutiny by research participants and audience, and if they present challenges to taken-for-granted ways of seeing and knowing, and provoke and promote questions about 'place' in the world, then perhaps this goes some way towards establishing the 'validity' of a qualitative research text. Communicating qualitative research is as much about how we know, as it is about what we know.

CONCLUSION

We have discussed the importance of language in the social construction of knowledge, how power is articulated through dichotomies and how meaning is inscribed in language. We believe models of writing which construct the writer as a disembodied narrator are inappropriate for communicating qualitative research. Our focus has been on written rather than visual texts, as writing remains the predominant means of communicating qualitative research. The breaking down of dichotomies, through an interpretative understanding of writing and researching as mutually constitutive processes, and an understanding of what principles might guide valid qualitative research are critical to writing 'good' qualitative research. It is also crucial to understand how power and meaning are inscribed in the words that constitute the research process, to recognise researchers' subjectivities, standpoint and locatedness (shifting and partial though they might be) and to acknowledge the voices of those with whom we undertake research. We believe that doing this enables a qualitative researcher to have confidence in the 'validity' of their interpretations. Consequently this chapter has not been a 'how to' guide, but a means of raising important issues that are inherent in the writing-IN process.

KEY TERMS

dichotomy
hegemony
modality
neo-positivism
nominalisation
objective
positionality
positivism
post-positivism
poststructuralism
reflexivity
situated knowledge
staging
subjective
trope
validity
writing-IN

REVIEW QUESTIONS

1. What implications do poststructuralist perspectives have for writing qualitative research?
2. In what ways is the term 'writing-up' misleading?
3. Why should writing be seen as an integral part of the entire research process?
4. How can a researcher endeavour to produce 'trustworthy' research?

SUGGESTED READING

Baxter, J. and Eyles, J. 1997, 'Evaluating qualitative research in social geography: establishing "rigour" in interview analysis', *Transactions of the Institute of British Geographers*, NS 22, pp. 505–525.

Bondi, L. 1997, 'In whose words? On gender identities, knowledge and writing practices', *Transactions of the Institute of British Geographers*, NS 22, pp. 245–58.

Jones, A. 1992, 'Writing Feminist Educational Research: Am "I" in the Text?', in *Women and Education in Aotearoa*, eds S. Middleton and A. Jones, Bridget Williams Books, Wellington.

McNeill, D. 1998, 'Writing the new Barcelona', in *The Entrepreneuurial City: Geographies of Politics, Regime and Representation*, eds T. Hall and P. Hubbard, John Wiley, London.

Richardson, L. 1994, 'Writing. A Method of Inquiry', in *Handbook of Qualitative Research*, eds N.K. Denzin and Y.S. Lincoln, Sage, Thousand Oaks.

Rose, G. 1997, 'Situating knowledges: Positionality, reflexivities and other tactics', *Progress in Human Geography*, vol. 21, pp. 305–320.

Glossary

accidental sampling—See *convenience sampling.*

aide mémoire—A list of topics to be discussed in an interview. May contain some clearly worded questions or key concepts intended to guide the interviewer. Alternative term for *interview guide.*

analytical log—Critical reflection on substantive issues arising in an interview. Links are made between emergent themes and the established literature or theory. (See also *personal log.*)

archival research—Research based on documentary sources (for example, public archives, photographs, newspapers).

asymmetrical power relation—A research situation in which informants are in positions of influence relative to the researcher. (See also *potentially exploitative power relation* and *reciprocal power relation.*)

attribute database—A set of information compiled from measurable characteristics, such as the census.

bias—Systematic error or distortion in a data set that might emerge as a result of researcher prejudices or methodological characteristics (for example, case selection, non-response, question wording, interviewer attitude).

bulletin board—An electronic medium devoted to sending and receiving messages for a particular interest group (for example, the about.com Geography Bulletin Board at <http://geography.about.com/education/scilife/geography/mpboards.htm> [accessed 4 November 1999]).

canon—A body of work, such as texts, held by some critics to be the most important of their kind and therefore worthy of serious study by all interested in the field.

case—Example of a more general process or structure that can be theorised. (See also *case study*.)

case study—Intensive study of an individual, group or place over a period of time. Research is typically done *in situ*.

CAQDA—Acronym for Computer Assisted Qualitative Data Analysis.

CATI—See *Computer-assisted telephone interviewing*.

chain sampling—See *snowball sampling*.

code and retrieve software—Specialist software packages that allow text to be segmented or grouped for coding and display. They facilitate electronic marking up, cutting, sorting, reorganising and collecting traditionally done with scissors, paper and sticky tape. Examples include *Kwalitan*™ and *The Ethnograph*™.

coding—Refers to the process of allocating information to particular categories in a form that then makes that data easy to retrieve and interpret. Coding may be text-based or numeric.

complete observation—A situation in which observation is overwhelmingly one-way and the researcher's presence is masked such that s/he is shielded from participation.

complete participation—A situation in which the researcher's immersion in a social context is such that s/he is first and foremost a participant. As a result of this level of immersion, the researcher may need to adopt critical distance to achieve an observational stance. That critical distance might be gained by reflection out-of-hours in the field, or through short-term exits from the field.

computer assisted cartography—Any hardware or software that is used to facilitate map making. *GIS* systems belong under this heading.

computer assisted qualitative data analysis software [CAQDAS]—Both a general acronym and the specific acronym for the CAQDAS network based in Surrey, United Kingdom.

computer-assisted telephone interviewing (CATI)—Questionnaire/interview conducted by telephone with questions being read directly from a computer file and responses being recorded directly onto a computer file.

concept building—Refers to the process of entering and coding data in a systematic way that relates to the research question being asked. The ability of a software package to support the systematic organisation of concepts leads to that software being categorised as *theory building software*.

conceptual mapping—As for *concept building*, but refers to the specific ability to visually represent data in some form. For example, *QSR NUD*IST*™ software uses hierarchical tree structures and ATLAS/*ti*™ uses network diagrams. *Inspiration*™ and *Decision Explorer*™ are purpose-built conceptual mapping programs for qualitative research.

confederate—Someone thought by other study participants to be another participant but who is, in fact, part of the research team.

confirmability—Extent to which results are shaped by respondents and not by researcher's *biases*.

content analysis—See *latent content analysis* and *manifest content analysis*.

controlled observation—Purposeful watching of worldly phenomena that is strictly limited by prior decisions in terms of scope, style and timing. (Compare with *uncontrolled observation*.)

convenience sampling—Involves selecting cases or participants on the basis of expedience. While the approach may appear to save time, money and effort, it is unlikely to yield useful information. Not recommended as a *purposive sampling* strategy.

credibility—The plausibility of an interpretation or account of experience.

critical inner dialogue—Constant attention to what the informant is saying, including *in situ* analysis of the themes being raised, and a continual assessment of whether the researcher fully understands what is being said.

critical reflexivity—See *reflexivity*.

criterion sampling—Choosing all cases that satisfy some predetermined standard.

'culturally safe'—Having knowledge of the history and beliefs and practices of minority groups and maintaining awareness of these factors.

data cleaning—Identifying and correcting errors of coding in a data set.

data management software—A generic term for any software which facilitates the entry, organisation, retrieval and/or coding and mapping of input data.

data retrieval—Refers to the process of getting access to data that has already been entered into the computer system.

data storage—Refers to the process of introducing data into a computer system so that it may be archived in some way (for example, as a document, spreadsheet, or graphics image).

debriefing—Procedure by which information about a research project (some of which may have been withheld or misrepresented) is made known to participants once the research is complete.

decode—To analyse in order to understand the hidden meanings in a *text*.

deconstruction—A method for challenging assumptions of coherence and truth within a *text* by revealing inconsistencies, contradictions and inadequacies (for example, where matters that are problematical have been naturalised).

deduction—Reasoning from principles to facts. (Compare with *induction*).

dependability—Minimisation of variability in interpretations of information gathered through research. Focuses attention on the researcher-as-instrument and the extent to which interpretations are made consistently.

dependent variable—A study item whose characteristics are considered to be influenced by an *independent variable*. For example flooding is heavily dependent on rainfall.

deviant case sampling—Selection of extraordinary cases (for example, outstanding successes, notable failures) to illuminate an issue or process of interest.

dichotomy—A division or binary classification in which one part of the dichotomy exists in opposition to the other (for example, light/dark, rich/poor). In most dichotomous thinking, one part of the binary is also more positively valued than the other.

disclosure—When a researcher reveals information about her/himself or the research project, or when research participants reveal information about themselves.

disconfirming case—Example that contradicts or calls into question researchers' interpretations and portrayals of an issue or process.

discourse—A system of signs through which realities are reproduced and legitimated. A discourse is a constructed system of arguments, ideologies and interpretations that shapes social practices, affecting the way we see things and talk about them.

eclecticism—An approach characterised by extensive borrowing of ideas from different *discourses* and their incorporation into a single argument.

epistemology–Ways of knowing the world and justifying belief. (See also ontology.)

essentialism—The idea that words (language) have some clear, apparent and fundamental meaning, rather than being given a meaning by users (readers, writers).

ethics—Refers to the moral conduct of researchers and their responsibilities and obligations to those involved in the research.

ethnography—A research method dependent on direct field observation in which the researcher is involved closely with a social group or neighbourhood. Also an account of events that occur within the life of a group, paying special attention to social structures, behaviour and the meaning(s) of these for the group.

extreme case sampling—See *deviant case sampling.*

facilitator—The person who encourages or moderates the discussion in a *focus group.*

factoid—A 'fact' which is considered to be of dubious origins or accuracy.

faction—A blend of 'fact' and 'fiction'. Usually refers to the tendency to weave interesting imaginary elements (for example, dialogue) into an otherwise factual account.

FAQ—Acronym for Frequently Asked Question.

fieldnotes—An accumulated written record of the fieldwork experience. May comprise observations and personal reflections. (Compare with *research diary*.)

focus group—A research method involving a small group of between six and ten people discussing a topic or issue defined by a researcher, with the researcher facilitating the discussion.

focused interview—Technique in which an interviewer poses a few predetermined questions but has flexibility in asking follow-up questions.

funnelling—Interview question ordering such that the topics covered move from general issues to specific or personal matters. (Compare with *pyramid structure*.)

generalisability—The degree to which research results can be extrapolated to a wider population group than that studied.

geographic information system (GIS)—A generic title for a number of integrated computer tools for the processing and analysing of geographical data, including specialized software for input (digitising) and output (printing or plotting) of mappable data. GIS is not the name of a specific software package.

GIS—See *Geographic Information System*.

graphic mapping—The graphical representation of concepts, often using nodes and links; conceptual network builders have graphic mapping capacity.

graphics software—Any computer program for the display and manipulation of pictures.

haptical quality—The immediate impact on the senses of a text, such as a film.

hegemony—A social condition in which people from all sorts of social backgrounds and classes come to interpret their own interests and consciousness in terms of the discourse of the dominant or ruling group. The hegemony of the dominant class is thus based, in part at least, on the (unwitting) consent of the subordinate classes. Such consent is created and reconstituted through the web of social relations, institutions, and public ideas in a society.

hermeneutics—The study of the interpretation of meaning in texts: whether there is assumed to be a single dominant meaning, or a multiplicity of meanings.

hierarchical tree building—The system for graphically representing coded and categorised data in a software program such as *NUD*IST*.

iconology—A search for deep symbolical images which provide a representation of the values of an individual or a group.

iconography—The identification and description of symbols and images.

image processing—Refers to the digital manipulations to which images are subjected in electronic systems.

independent variable—A study item whose characteristics are considered to cause change in a dependent variable. For example the independent variable rainfall may promote flooding (the *dependent variable*).

induction—Process of generalisation involving the application of specific information to a general situation or to future events. (Compare with *deduction*.)

informant—Person interviewed by a researcher. Some refer to those who are interviewed as 'subjects' or 'respondents'. Others argue that someone who is interviewed, as opposed to simply observed or surveyed, is more appropriately referred to as an informant. That is because an interview informant is likely to have a more active and informed role in the research encounter.

informed consent—Informant/subject agreement to participate in a study having been fully apprised of the conditions associated with that study (for example, time involved, methods of investigation, likely inconveniences and possible consequences).

insider—A research position in which the researcher is socially accepted as being 'inside', or a part of, the social groups or places involved in the study. (Compare with *outsider*.)

interpretive community—Involves established disciplines with relatively defined and stable areas of interest, theory, and research methods and techniques. Influences researchers' choice of topics and approaches to, and conduct of, study.

intersubjectivity—Meanings and interpretations of the world created, confirmed or disconfirmed as a result of interactions (language and action) with other people within specific contexts. (See also *subjectivity* and *objectivity*.)

intertextuality—The necessary interdependence of a *text* with those that have preceded it. Any text is built upon and made meaningful by its associations with others.

interview—A means of data collection involving an oral exchange of information between the researcher and one or more other people.

interview guide—A list of topics to be covered in an *interview*. May contain some clearly worded questions or key concepts intended to guide the interviewer. (Compare with *interview schedule*.)

interview schedule—Ordered list of questions which the researcher intends to ask informants. Questions are worded similarly and are asked in the same order for each informant. In its most rigid form, an interview schedule is a questionnaire delivered in face-to-face format. (Compare with *interview guide*.)

landscape—Landscape is used broadly to mean a built, cultural or physical environment (and even the human body) which can be 'read' and interpreted.

latent content analysis—Assessment of implicit themes within a text. Latent content may include ideologies, beliefs, or stereotypes. (Compare with *manifest content analysis*.)

life history—An *interview* in which data on the experiences and events of a person's life are collected. The aim is to gain insights into how a person's life may have been affected by institutions, social structures, relations, rites of passage or other significant events. (Compare with *oral history*.)

manifest content analysis—Assessment of the surface or visible content of text. Visible content may include specific words, phrases or the physical space dedicated to a theme (for example, column centimetres in a newspaper). (Compare with *latent content analysis*.)

margin-coding—A simple system of categorising material in *transcripts*. Typically involves marking the transcript margin with a colour, number, letter or symbol code to represent key themes or categories.

maximum variation sampling—Form of sampling based on high diversity aiming to uncover systematic variations and common patterns within those variations.

memoing—Is a process which has specific meaning in qualitative software systems: the researcher may write memos or reflections on the research process as s/he works and then incorporate these memos as electronic data for further investigation.

metaphor—An expression applied to something to which it is not literally applicable in order to highlight an essential characteristic.

mimetic—Miming or imitating.

mixed methods—A combination of techniques for tackling a research problem; the term is often used specifically to mean a combination of quantitative and qualitative methods.

moderator—See *facilitator*.

multiple voices—A reference to the need to listen to alternative literatures, *texts*, expressions or opinions, and therefore avoid the assumption that there is only one view of merit.

nominalisation—The transformation of verbs and adjectives into nouns. Nominalisation reduces information available to readers and it mystifies social processes by hiding actions and the identity of actors.

NUD*IST—A software system for qualitative data analysis developed by Richards and Richards at Qualitative Research Solutions in Australia. The acronym stands for Nonnumerical, Unstructured Data: Indexing, Searching, Theorising.

objective/objectivity—Unaffected by feelings, opinions or personal characteristics. Often contrasted with *subjectivity*. (See also *intersubjectivity*.)

objective modality—A form of writing that implicitly hides the writer's presence in the text (e.g. third-person narrative form) but which clearly signals agreement with the statement being made. (Compare with *subjective modality*.)

observation—Most literally, purposefully watching worldly phenomena. Increasingly broadened beyond seeing to include apprehending the environment through all our senses (for example, sound, smell) for research purposes.

observer-as-participant—A research situation in which the researcher is primarily able to observe but in so doing is also participating in a social situation. (Compare with *participant-as-observer*.)

ontology—beliefs about the world. Understanding about the kinds of things that exist in the universe and the relations between them. (See also *epistemology*.)

opportunistic sampling—Impromptu decision to involve *cases* or *participants* in a study on the basis of leads uncovered during fieldwork.

oral history—Interviews which collect data about historical events, spaces and periods. Often used to collect data that adds to, or fill gaps in, existing historical records. (Compare with *life history*.)

oral methods—Verbal techniques, such as interviews or focus groups, as opposed to written methods for seeking information.

'other'—An outside or different subject against which a person compares and establishes their own social position, meaning and identity. Also taken to mean that which is oppositional to the mainstream, marginal, or outside the dominant ideology.

outsider—A research position in which the researcher is rendered 'outside' a social circle, or feels 'out of place' on account of differences such as visible appearance, unfamiliarity, or inability to speak the language or vernacular used. (Compare with *insider*.)

panopticon—A circular prison with cells surrounding a central warders' station. In the panopticon, inmates may be observed at any time but they are not aware of the observation.

participant—Person taking part in a research project. Usually the *informant*, rather than a member of the research team.

participant checking—Informant's review of the *transcript* of their contribution to an interview or focus group for accuracy and meaning. May also involve informant reviewing the overall research output (for example, thesis, report). Also serves as a means of continuing the involvement of informants in the research process.

participant-as-observer—A research situation in which the researcher is primarily a participant in a social situation or gathering place, but in so doing can maintain sufficient critical distance to observe social dynamics and interactions. (Compare with *observer-as-participant*.)

participant observation—A fieldwork method in which the researcher studies a social group while being a part of that group.

pastiche—A *text* which is a medley drawn from various sources.

personal log—Recorded reflections on the practice of an interview. Includes discussions of the appropriateness of the order and phrasing of questions, and of the informant selection. Also contains assessments of matters such as research design and ethical issues. (See also *analytical log.*)

pilot study—Abbreviated version of a research project in which the researcher practices or tests procedures to be used in a subsequent full-scale project.

positionality—A researcher's social, locational and ideological placement relative to the research project or to other participants in it. May be influenced by biographical characteristics such as class, race and gender, as well as various formative experiences.

positivism—An approach to scientific knowledge based around foundational statements about what constitutes truth and legitimate ways of knowing. There is a number of variants of positivist thought but central to all is the construction a singular universal and value-free knowledge based on empirical observation and the scientific method.

postcolonialism—An approach to knowledge that seeks to represent voices of the '*other*', especially colonised peoples and women, and to recognise knowledge that has been ignored through processes of colonisation and patriarchy.

poststructuralism—A school of thought which endeavours to link language, subjectivity, social organisation and power.

potentially exploitative power relation—A research situation in which the researcher is in a position of power relative to the investigator. (See also *asymmetrical power relation* and *reciprocal power relation.*)

primary observation—Research in which the investigator is a participant in, and interpreter of, human activity involving her/his own experience. (Compare with *secondary observation.*)

primary question—Interview question used to initiate discussion of a new topic or theme. (Compare with *probe question.*)

probability sampling—Sampling technique intended to ensure a random and statistically representative sample that will allow confident generalisation to the larger population from which the sample was drawn. (Compare with *purposive sampling*).

probe question—A gesture or follow-up question used in an interview to explore further a theme or topic already being discussed. (See also *prompt*. Compare with *primary question.*)

prompt—A follow-up question in an interview designed to deepen a response (for example, 'why do you say that?'; 'what do you mean?'.) (See also *probe question*. Compare with *primary question.*)

purposive sampling—Sampling procedure intended to obtain a particular group for study on the basis of specific characteristics they possess. (Compare with *probability sampling.*) Aims to uncover information-rich phenomena/participants that can shed light on issues of central importance to the study.

purposeful sampling—See *purposive sampling.*

pyramid structure—Order of interview questions in which easy-to-answer questions are posed at the beginning of the interview while deeper or more philosophical questions/issues are raised at the end. (Compare with *funnelling.*)

quantitative methods—Statisical and mathematical modelling approaches used to understand social and physical relationships.

quota sampling—Selecting sampling elements on the basis of categories known or assumed to exist in the universe population.

random sampling—See *probability sampling.*

rapport—A productive interpersonal climate between informant and researcher. A relationship that allows the informant to feel comfortable or confident enough to offer comprehensive answers to questions.

reactive effects—The influence a research method or researcher has on the individuals or phenomena under observation.

reciprocal power relation—A research situation in which researcher and informant are in comparable social positions and experience relatively equal costs and benefits of participating in the research. (See also *asymmetrical power relation* and *potentially exploitative power relation.*)

recruitment—The process of finding people willing to participate in a research project. Recruitment strategies can range from asking people 'on the street' (perhaps to fill in a questionnaire) to inviting key individuals to participate (in a focus group, for example).

reflexivity—Self-critical introspection and a self-conscious scrutiny of oneself as a researcher.

reliability—Extent to which a method of data collection yields consistent and reproducible results when used in similar circumstances by different researchers or at different times. (See also *validity*.)

relativism—An approach to knowledge in which it is held that there is no means for significantly differentiating between the merits of arguments. Suggests that there are no absolute, unequivocal standards of true/false or right/wrong.

replicability—Able to be repeated or tested to see how general the particular findings of a study are in the wider population.

representation—The way in which something (the world, human behaviour, a city, the landscape) is depicted, recognising that this cannot be an exact depiction. An important insight from *poststructuralist* thinkers is that representations not only describe the social world but also help to shape or constitute it.

research diary—A place for recording observations in the process of being critically reflexive. Contains thoughts and ideas about the research process, its social context and the researcher's role in it. The contents of a research diary are different to those of the *fieldnotes*, which more typically contain qualitative data, such as records of observations, conversations and sketch maps.

respondent—See *informant*.

rigour—Accuracy, exactitude and trustworthiness.

sample—Phenomena or participants selected from a larger set of phenomena or a larger population for inclusion in a study.

sampling—Means of selecting phenomena or participants for inclusion in a study. A key difference between qualitative and quantitative enquiry is in the logic underpinning their use of *purposive* and *probability (random) sampling* respectively.

secondary observation—Research in which the data are the observations of others (for example, photographs).

semiology—See *semiotics.*

semiotics—The system or language of *signs* (sometimes referred to as semiology). (See also *signifier* and *signified.*)

semi-structured interview—Interview with some predetermined order, but which nonetheless has flexibility with regard to the position/timing of questions. Some questions, particularly sensitive or complex ones, may have a standard wording for each informant. (Compare with *structured interview* and *unstructured interview.*)

sign—Written or other image/mark that represents something else. Comprises *signifier* and *signified.*

signified—The meaning derived from a *signifier* (or from a set of signifiers, such as a *text*).

signifier—Images such as written marks, or features of the landscape with which meaning is associated.

situated knowledge—A metaphor that evokes recognition of the positionality (or contextual nature) of knowledges. The inscription and creation of knowledge is always partial and 'located' somewhere.

snowball sampling—Sampling technique that involves finding participants for a research project by asking existing informants to recommend others who might be interested. From one or two participants the number of people involved in the project 'snowballs'. Also known as chain sampling.

social structure—See *structure.*

staging—A theatrical metaphor for writing-in research that encourages geographical researchers to consider how the construction of a research text is actually a form of cultural production. The author of the text is a creator, director and performer in the particular narrative he or she is constructing.

standardised questions—A uniform set of questions that are repeated for all interviews or focus groups in a research project, contrasting with spontaneous questions that develop out of the conversational flow of an interview or focus group.

structure—A structure is a functioning system (for example, social, economic or political) within which individuals are located, within which all events are enacted, and which are reproduced and transformed by those events.

structured interview—Interview that follows a strict order of topics. Usually the order is set out in an *interview schedule*. The wording of questions for each interview may also be predetermined. (Compare with *unstructured interview* and *semi-structured interview.*)

subject—See *informant.*

subjective/subjectivity—Refers to the insertion of the personal resources, opinions and characteristics of a person into a research project. Often contrasted with *objectivity*. (See also *intersubjectivity.*)

subjective modality—A form of writing that explicitly acknowledges the writer's presence in the text (e.g., first-person narrative form) and clearly signals their agreement or disagreement with the statement being made. (Compare with *objective modality.*)

text—Traditionally synonymous with the written page, but now used more broadly to include a range of forms, such as music, paintings, photographs, maps, landscapes (and even institutions) as well as the written word.

text based managers—Software with capacity for managing and organising data, creating subsets of data for further analysis, and searching and retrieving combinations of words, phrases, coded segments, memos or other material. Examples include *askSam*™, *FolioVIEWS*™ and *winMAX*™.

text retriever software—Computer program for recovering data by category on the basis of keywords that appear in the data; for finding words, phrases or other character strings; and for finding things that are misspelt, sound alike, mean the same thing, or have certain patterns. Examples include *Metamorph*™, *The Text Collector*™, *WordCruncher*™, *ZyINDEX*™ and *Sonar Professional*™.

textual analysis—Reading and constant reinterpretation of *texts* as a set of *signs* or signifying practices.

textual community—Group of individuals who share certain understandings of the meaning of *texts.*

theory building software—Computer programs that deal with relationships between data categories to develop higher-order classifications and categories, and to formulate and test propositions or assertions. Examples include *AQUAD™*, *ATLAS/ti™*, *HyperRESEARCH™* and *NUD*IST™*.

transcript—Written record of speech (for example, interview, focus group proceedings, film dialogue). May also include textual description of informant gestures and tone.

transferability—Extent to which the results of a study might apply to contexts other than that of the research study.

triangulation—Use of multiple or mixed methods, researchers and information sources to confirm or corroborate results.

trope—Figure of speech that allows writers or producers of other forms of text to say one thing but mean something else. May involve use of metaphor or metonymy.

typical case sampling—Selection of samples that illustrate or highlight that which is considered typical or normal.

uncontrolled observation—Purposeful watching of worldly phenomena that is relatively unconstrained by restrictions of scope, style and time. (Compare with *controlled observation*.)

unstructured interview—Interview in which there is no predetermined order to the issues addressed. The researcher phrases and raises questions in a manner appropriate to the informants' previous comment. The direction and vernacular of the interview is informant driven. (Compare with *structured interview* and *semi-structured interview*.)

validity—The truthfulness or accuracy of data compared with acceptable criteria. (See also *reliability*.)

vernacular—Occurring in the location where it originated. Vernacular language is the native language of a place.

warm-up—A set of pre-interview techniques intended to enhance rapport between interviewer and informant. May include small talk, sharing food, or relaxed discussion of the research.

word processing—A generic concept which includes the use of computer capacity to create, edit, and print documents.

word searching—The process of looking for individual words in an electronic text.

References

Abraham, F. 1982, *Modern Sociological Theory: An Introduction*, Oxford University Press, New Delhi.

Adelman, C. (ed.) 1981, *Uttering, Muttering: Collecting, Using and Reporting Talk for Social and Educational Research*, Grant McIntyre, London.

Agar, M. 1986, *Speaking of Ethnography*, Sage, Beverly Hills.

Agar, M. and MacDonald, J. 1995, 'Focus groups and ethnography', *Human Organization*, vol. 54, no. 1, pp. 78–86.

Aitken, S. C. and Zonn, L. E. 1993, 'Wier(d) sex: representation of gender-environment relations in Peter Weir's *Picnic at Hanging Rock* and *Gallipoli*', *Environment and Planning D: Society and Space*, vol. 11, no. 2, pp. 191–212.

Alder, P. A. and Alder, P. 1994, 'Observational techniques', in *Handbook of Qualitative Research*, (eds) N. K. Denzin and Y. S. Lincoln, Sage, Thousand Oaks.

Alvermann, D. E., O'Brien, D. G. and Dillon, D. R. 1996, 'On writing qualitative research', *Reading Research Quarterly*, vol. 31, pp. 114–20.

Anderson, K. J. 1999, 'Reflections on Redfern', in E. Stratford (ed.), *Australian Cultural Geographies*, Oxford University Press, Melbourne.

— 1995, 'Culture and nature at the Adelaide Zoo: at the frontiers of "human" geography', *Transactions of the Institute of British Geographers*, vol. 20, no. 3, pp. 275–94.

— 1993, 'Place narratives and the origins of inner Sydney's Aboriginal settlement, 1972–73', *Journal of Historical Geography*, vol. 19, no. 3, pp. 314–35.

Anderson, K. J. and Gale, F. 1992, *Inventing Places: Studies in Cultural Geography*, Longman Cheshire, Melbourne.

Anzaldúa, G. 1987, *Borderlands/La Frontera: The New Mestiza*, Spinsters/Aunt Lute Press, San Francisco.

Atkinson, P. and Hammersley, M. 1984, 'Ethnography and participant observation', in *Handbook of Qualitative Research*, (eds) N. K. Denzin and Y. S. Lincoln, Sage, Thousand Oaks.

Ayres, L. 1997, Defining and Managing Family Caregiving in Chronic Illness: Expectations, Explanations and Strategies, PhD thesis, University of Illinois at Chicago.

Babbie, E. 1998. *The Practice of Social Research*, 8th edn, Wadsworth Publishing Company, Belmont.

— 1992, *The Practice of Social Research*, 6th edn, Wadsworth Publishing Company, Belmont.

Bailey, C., White, C. and Pain, R. 1999a, 'Evaluating qualitative research: dealing with the tension between "science" and "creativity"', *Area*, vol. 31, no. 2, pp. 169–83.

— 1999b, 'Response', *Area* vol. 31, no. 2, pp. 183–4.

Barnes, T. J. 1993, 'Whatever happened to the philosophy of science?', *Environment and Planning* A, vol. 25, pp. 301–04.

— 1989, 'Place, space and theories of economic value: Contextualism and essentialism', *Transactions of the Institute of British Geographers*, NS 14, pp. 299–316.

Barnes, T. and Duncan, J. 1992a, 'Introduction: writing worlds', in *Writing Worlds: Discourse, Text and Metaphor in the Representation of Landscape*, (eds) T. Barnes and J. Duncan, Routledge, London.

— (eds) 1992b, *Writing worlds: Discourse, text and metaphor in the representation of landscape*, Routledge, London.

Barnes, T. J. and Gregory, D. 1997, 'Worlding geography: Geography as situated knowledge', in *Reading Human Geography: The poetics and politics of inquiry*, (eds) T. J Barnes and D. Gregory, Arnold, London.

Barry, C. 1998, 'Choosing Qualitative Data Analysis Software: Atlas/ti and Nudist Compared', *Sociological Research Online*, vol. 3, no. 3, <http://www.socresonline.org.uk/socresonline/3/3/4.html> (8 February 1999).

Baxter, J. 1998, Exploring the Meaning of Risk and Uncertainty in an Environmentally Sensitized Community, PhD thesis, Department of Geography, McMaster University, Ontario.

Baxter, J. and Eyles, J. 1999a, 'The utility of in-depth interviews for studying the meaning of environmental risk', *Professional Geographer*, vol. 51, no. 2, pp. 307–20.

— 1999b, 'Prescription for research practice? Grounded theory in qualitative evaluation', *Area*, vol. 31, no. 2, pp. 179–81

— 1997, 'Evaluating qualitative research in social geography: establishing "rigour" in interview analysis', *Transactions of the Institute of British Geographers*, vol. 22, no. 4, pp. 505–25.

Bazely, P. 1997, 'NUD*IST 4: — Survey research: How do I set up data input for a survey to link structured responses (e.g. in SPSS) with qualitative data in NUD.IST 4?', *User Support Notes*. <http://www.qsr.com.au/software/n4/usn/usn4017.htm> (8 February 1999).

Bell, D. 1991, 'Art and land in New Zealand', *New Zealand Journal of Geography*, vol. 92, pp. 15–17.

Bell, D., Binnie, J., Cream, J. and Valentine, G. 1994, 'All hyped up and no place to go', *Gender Place and Culture*, vol. 1, no. 1, pp. 31–48.

Bell, D., Caplan, P. and Karim, W. J. 1993, *Gendered Fields: Women, Men and Ethnography*, Routledge, London.

Berg, B. L. 1989, *Qualitative Research Methods for the Social Sciences*, Allyn and Bacon, Boston.

Berg, L. D. 1997, 'Banal Geographies', paper presented to the Inaugural International Conference of Critical Geographers, Vancouver, Canada, August 10–13, 1997. Available online on the World Wide Web at: <http://www.geog.ubc.ca/iiccg/papers/Berg_L.html>.

— 1994a, 'Masculinity, Place, and a Binary Discourse of Theory and Empirical Investigation in the Human Geography of Aotearoa/New Zealand', *Gender, Place and Culture*, vol. 1, no. 2, pp. 245–60.

— 1994b, 'Masculinism, power and discourses of exclusion in Brian Berry's "'Scientific" Geography', *Urban Geography*, vol. 15, pp. 279–87.

— 1993, 'Between Modernism and Postmodernism', *Progress in Human Geography*, vol. 17, pp. 490–507.

Berg, L. D. and Kearns, R. A. 1998, 'America Unlimited', *Environment and Planning D: Society and Space*, vol. 16, pp. 128–32.

Berg, L. and Mansvelt, J. 2000, 'Writing in, speaking out: communicating qualitative research findings', in I. Hay (ed.), *Qualitative Research Methods in Human Geography*, Oxford University Press, Melbourne.

Bernard, H. R. 1988, *Research Methods in Cultural Anthropology*, Sage, Newbury Park.

Bertrand, J. T., Brown, J. E. and Ward, V. M. 1992, 'Techniques for analyzing focus group data', *Evaluation Review*, vol. 16, no. 2, pp. 198–209.

Billinge, M., Gregory, D. and Martin, R. 1984, 'Reconstructions', in M. Billinge, D. Gregory and R. L. Martin (eds), *Recollections of a Revolution: Geography as Spatial Science*, Macmillan, London.

Bogdan, R. 1974, *Being Different: The Autobiography of Jane Frey*, Wiley, London.

Bogdan, R. C. and Biklen, S. K. 1992, *Qualitative Research for Education: An Introduction to Theory and Methods*, 2nd edn, Allyn and Bacon, Boston.

Bondi, L. 1997, 'In whose words? On gender identities, knowledge and writing practices', *Transactions of the Institute of British Geographers*, vol. 22, pp. 245–58.

Bordo, S. 1986, 'The Cartesian masculinization of thought', *Signs*, vol. 11, pp. 439–56.

Bouma, G. D. 1996, *The Research Process*, 3rd edn, Oxford University Press, Melbourne.

— 1993, *The Research Process*, rev. edn, Oxford University Press, Melbourne.

Brannen, J. (ed.) 1992a, *Mixing Methods: Qualitative and Quantitative Research*, Avebury, Aldershot.

— 1992b, 'Combining qualitative and quantitative approaches: an overview', in J. Brannen (ed.), *Mixing Methods: Qualitative and Quantitative Research*, Avebury, Aldershot.

Bryman, A. 1984, 'The debate about quantitative and qualitative research: a question of method or epistemology?', *The British Journal of Sociology*, vol. 35, pp. 75–92.

Bryman, A. and Burgess, R. G. (eds) 1994, *Analyzing Qualitative Data*, Routledge, London.

Burgess, J. 1996, 'Focusing on fear: The use of focus groups in a project for the Community Forest Unit, Countryside Commission', *Area*, vol. 28. no. 2, pp. 130–35.

— 1988, 'Exploring environmental values through the medium of small groups: 2. Illustrations of a group at work', *Environment and Planning* A, vol. 20, no. 4, pp. 457–76.

Burgess, J., Limb, C. and Harrison, C. M. 1988, 'Exploring environmental values through the medium of small groups: 1. Theory and practice', *Environment and Planning* A, vol. 20, no. 3, pp. 309–26.

Burgess, J. and Wood, P. 1988, 'Decoding docklands: place advertising and the decision-making strategies of the small firm', in J. Eyles and D. M. Smith (eds), *Qualitative Methods in Human Geography*, Polity Press, Cambridge.

Burgess, R. G. 1996, (ed.) *Studies in Qualitative Methodology: Computing and Qualitative Research:* Vol. 5, JAI Press, London.

— 1982a, 'Elements of sampling in field research', in R. G. Burgess (ed.), *Field Research: A Sourcebook and Field Manual*, Allen and Unwin, London.

— 1982b, 'Multiple strategies in field research', in R. G. Burgess (ed.), *Field Research: A Sourcebook and Field Manual*, George Allen and Unwin, London.

— 1982, 'The unstructured interview as a conversation', in R. G. Burgess (ed.), *Field Research: A Sourcebook and Field Manual*, George Allen and Unwin, London.

Burawoy, M., Burton, A., Ferguson, A., Fox, K., Gamson, J., Gartrell, N., Hurst, L., Kurzman, C., Salzinger, L., Schiffman, J. and Ui, S. 1991, *Ethnography unbound: power and resistance in the modern metropolis*, University of California Press, Berkeley.

Buston, K. 1997, 'NUD*IST in Action: Its use and its Usefulness in a Study of Chronic Illness in Young People', *Sociological Research Online*, vol. 2, no. 3, <http://www.socresonline.org.uk/socresonline/2/3/6.html> (8 February 1999).

Butler, R. 1997, 'Stories and experiments in social inquiry', *Organisation Studies*, vol. 18, no. 6, pp. 927–48.

Cameron, J. 1992, Modern-Day Tales of Illegitimacy: Class, Gender and Ex-Nuptial Fertility, MA Minor Thesis, Department of Geography, University of Sydney.

Carey, M. A. 1994, 'The group effect in focus groups: Planning, implementing and interpreting focus group research', in J. M. Morse (ed.), *Critical Issues in Qualitative Research Methods*, Sage, Thousand Oaks.

Casey, M. J., Wyatt, G., Rager, A., Arnevik, C., Pfarr, D., Anderson, J., Everett, L. and Busman, L. 1996, *Addressing Nonpoint Source Agricultural Pollution in the Minnesota River Basin: Findings from Focus Groups Conducted with Farmers, Agency Staff, Crop Consultants and Researchers, December 1995*, Study Conducted for the Minnesota Department of Agriculture [on line], available at <http://www.soils.agri.umn.edu/research/mn-river/doc/fgrptweb.html>

Cavendish, A. P. 1964, 'Early Greek philosophy', in D. J. O'Connor (ed.), *A Critical History of Western Philosophy*, The Free Press, New York.

Ceglowski, D. 1997, 'That's a good story, but is it really research?', *Qualitative Inquiry*, vol. 3, no. 2, pp. 188–99.
Chatwin, B. 1987, *The Songlines*, Pan Books, London.
Clarke, D. B. 1997, 'Introduction: previewing the cinematic city', in D. B. Clarke (ed.), *The Cinematic City*, Routledge, London.
Collins, D. and Kearns, R. 1998, *Avoiding the Log-jam: Exotic Forestry, Transport and Health in Hokianga*, Working Paper No. 8, Department of Geography, The University of Auckland.
Connell, R. W. 1991, 'Live fast and die young: the construction of masculinity among young working-class men on the margin of the labour market', *Australia and New Zealand Journal of Sociology*, vol. 27, no. 2, pp. 141–71.
Cook, I. 1997, 'Participant observation', in R. Flowerdew and D. Martin (eds), *Methods in Human Geography*, Addison Wesley Longman, Harlow.
Cooke, K. 1997, 'Tragedy in one ACT', *The Australian Magazine*, 19–20 July, p 53.
Cooper, A. 1994, 'Negotiating dilemmas of landscape, place and Christian commitment in a Suffolk parish', *Transactions of the Institute of British Geographers*, vol. 19, pp. 202–12.
Cooper, A. 1995, 'Adolescent dilemmas of landscape, place and religious experience in a Suffolk parish', *Environment and Planning D: Society and Space*, vol. 13, pp. 349–63.
Crang, M. 1997a, 'Analyzing qualitative materials', in R. Flowerdew and D. Martin (eds), *Methods in Human Geography: A Guide for Students Doing Research*, Addison Wesley Longman, Harlow.
— 1997b, 'Picturing practices: research through the tourist gaze', *Progress in Human Geography*, vol. 21, pp. 359–73.
Crang, M., Hudson, A., Reimer, S., and Hinchliffe, S. 1997, 'Software for Qualitative Research: 1. Prospectus and overview', *Environment and Planning A*, vol. 29, no. 5, pp. 771–87.
Crang, P. 1996, 'It's showtime: on the workplace geographies of display in a restaurant in southeast England', *Environment and Planning D: Society and Space*, vol. 12, pp. 675–704.
Crush, J. 1993, 'Post-colonialism, de-colonization, and geography', in A. Godlewska and N. Smith (eds), *Geography and Empire*, Blackwell, Oxford.
Daniels, S. and Cosgrove, D. 1988, 'Introduction: iconography and landscape', in D. Cosgrove and S. Daniels (eds), *The Iconography of Landscape: Essays on the Symbols, Representation, Design and Use of Past Environments*, Cambridge University Press, Cambridge.
Davis, C. M. 1954, 'Field techniques', in P. E. James and C. F. Jones (eds), *American Geography: Inventory and Prospect*, Syracuse University Press for the AAG, Syracuse.
Davis, M. 1997, *Gangland: Cultural Elites and the New Generationalism*, Allen and Unwin, Sydney.
— 1990, *City of Quartz: Excavating the future in Los Angeles*, Verso, London.
Dear, M. 1988, 'The postmodern challenge. Reconstructing human geography', *Transactions of the Institute of British Geographers*, NS 13, pp. 262–74.
Denzin, N. K. 1994, 'The Art and Politics of Interpretation', in N. K. Denzin and

Y. S. Lincoln (eds), *Handbook of Qualitative Research*, Sage, Thousand Oaks.
— 1978, *The Research Act*, 2nd edn, McGraw Hill, New York.
Denzin, N. and Lincoln, Y. 1994, 'Introduction: entering the field of qualitative research.' in N. Denzin and Y. Lincoln (eds), *Handbook of Qualitative Research*, Sage, Thousand Oaks.
Derrida, J. 1981, *Dissemination*, translated by B. Johnson, University of Chicago Press, Chicago.
— 1978, *Writing and Difference*, translated with an introduction and additional notes by A. Bass, University of Chicago Press, Chicago.
— 1976, *Of Grammatology*, translated by G. Spivak, The Johns Hopkins University Press, Baltimore and London.
Dey, I. 1993, *Qualitative Research Analysis: A User Friendly Guide for Social Scientists*, Routledge, London.
Dixon, D. P. and Jones III, J. P. 1996, 'For a Supercalifragilisticexpialidocious Scientific Geography', *Annals of the Association of American Geographers*, vol. 86, pp. 767–79.
Donovan, J. 1988, 'When you're ill, you've gotta carry it', in J. Eyles and D. M. Smith (eds), *Qualitative Methods in Human Geography*, Polity Press, Cambridge.
Douglas, J. D. 1985, *Creative Interviewing*, Sage, Beverly Hills.
Douglas, L., Roberts, A. and Thompson, R. 1988, *Oral History: A Handbook*, Allen and Unwin, Sydney.
Duncan, J. 1992, 'Elite landscapes as cultural (re) production: the case of Shaughnessy Heights', in K. Anderson and F. Gale (eds), *Inventing Places: Studies in Cultural Geography*, Longman Cheshire, Melbourne.
—1987, 'Review of urban imagery: urban semiotics', *Urban Geography*, vol. 8, no. 5, pp. 473–83.
Duncan, J. and Duncan, N. 1998, 'Re-reading the landscape', *Environment and Planning D: Society and Space*, vol. 6, no. 2, pp. 117–26.
Dunn, K. M. 1995, 'The landscape as text metaphor', in G. Dixon and D. Aitken (eds), *IAG Conference Proceedings 1993*, Monash Publications in Geography, Melbourne.
— 1993, 'The Vietnamese concentration in Cabramatta: Site of avoidance and deprivation, or island of adjustment and participation?', *Australian Geographical Studies*, vol. 31, no. 2, pp. 228–45.
Dunn, K. M., McGuirk, P. M. and Winchester, H. P. M. 1995, 'Place making: the social construction of Newcastle', *Australian Geographical Studies*, vol. 33, no. 2, pp. 149–66.
Dyck, I. 1999, 'Using qualitative methods in medical geography: deconstructive moments in a subdiscipline?', *Professional Geographer*, vol. 51, no. 2, pp. 243–53.
— 1997, 'Dialogue with difference: A tale of two studies', in J. P. Jones, III, H. L. Nast and S. M. Roberts (eds), *Thresholds in Feminist Geography: difference, methodology, representation*, Rowman and Littlefield, Lanham.
Dyck, I. and Kearns, R. 1995, 'Transforming the relations of research: towards culturally safe geographies of health and healing', *Health and Place*, vol. 1, no. 3, pp. 137–47.

Earickson, R. and Harlin, J. 1994, *Geographic Measurement and Quantitative Analysis*, Prentice Hall, Upper Saddle River.

Edwards, J. A. and Lampert, M. D. (eds) 1993, *Talking Data: Transcription and Coding in Discourse Research*, Lawrence Erlbaum Associates, Hillsdale, New Jersey.

Eichler, M. 1988, *Nonsexist Research Methods: A Practical Guide*, Allen and Unwin, Boston.

England, K. V. L. 1994, 'Getting personal: reflexivity, positionality, and feminist research', *Professional Geographer*, vol. 46, pp. 80–89.

— 1993, 'Suburban pink collar ghettos: the spatial entrapment of women', *Annals of the Association of American Geographers*, vol. 83, no. 2, pp. 225–42.

Evans, M. 1988, 'Participant observation: the researcher as research tool', in J. Eyles and D. M. Smith (eds), *Qualitative Methods in Human Geography*, Polity Press, Cambridge.

Eyles, J. 1988,· 'Interpreting the geographical world: qualitative approaches in geographical research', in J. Eyles and D. Smith (eds), *Qualitative Methods in Human Geography*, Polity Press, Cambridge.

Eyles, J. and Smith, D. (eds) 1988, *Qualitative Methods in Human Geography*, Polity Press, Cambridge.

Fairclough, N. 1992, *Discourse and Social Change*, Polity Press, Cambridge.

Feinsilver, J. M. 1993, *Healing the Masses: Cuban health politics at home and abroad*, University of Columbia Press, Berkeley.

Fielding, N. 1999, 'The norm and the text: Denzin and Lincoln's handbooks of qualitative method', *British Journal of Sociology*, vol. 50, no. 3, pp. 525–34.

— 1995, 'Choosing the right software program', *ESRC Data Archive Bulletin*, no. 58, <http://kennedy.soc.surrey.ac.uk/caqdas/choose.htm> (8 February 1999).

— 1994, 'Getting into computer-aided qualitative data analysis', *ESRC Data Archive Bulletin*, September, no. 57, <http://kennedy.soc.surrey.ac.uk/caqdas/getting.htm> (8 February 1999).

Fielding, N. and Lee, R. 1998, *Computer Analysis and Qualitative Research*, Sage, London.

Findlay, A. M. and Li, F. L. N. 1997, 'An auto-biographical approach to understanding migration: the case of Hong Kong emigrants', *Area*, vol. 29, no. 1, pp. 34–44.

Fink, A. and Kosecoff, J. 1985, *How to Conduct Surveys: A Step-by-Step Guide*. Sage Publications, Beverly Hills.

Fish, S. 1980, *Is There a Text in this Class? The Authority of Interpretive Communities*, Harvard University Press, London.

Flick, U. 1992, 'Triangulation revisited: strategy of validation or alternative?', *Journal for the Theory of Social Behaviour*, vol. 2, no. 2, pp. 175–97.

Flowerdew, R. and Martin, D. (eds) 1997, *Methods in Human Geography*, Addison Wesley Longman, Harlow.

Flyvbjerg, B. 1998, *Rationality and Power: Democracy in Practice*, translated by S. Sampson, University of Chicago Press, Chicago.

Forbes, D. K. 1999, 'Globalisation, postcolonialism and new representations of the Pacific Asian metropolis', in P. Dicken, P. Kelly, L. Kong, K. Olds and H. Yeung (eds), *The Logic(s) of Globalisation*, Routledge, London.

— 1993, 'Multiculturalism, the Asian connection and Canberra's urban imagery', in G. Clark, D. Forbes and R. Francis (eds), *Multiculturalism, Difference and Postmodernism*, Longman Cheshire, Melbourne.

Forer, P. and Chalmers, L. 1987, 'Geography and information technology: issues and impacts', in P. G. Holland and W. B. Johnston (eds), *Southern Approaches: Geography in New Zealand*, New Zealand Geographical Society (Inc.), Christchurch.

Foucault, M. 1977a, *Discipline and Punish: the birth of the prison*, Pantheon, New York.

— 1977b, *Language, counter-memory, practice: Selected essays and interviews*, Cornell University Press, Ithaca.

Fowler, R. 1991, *Language in the News: Discourse and ideology in the press*, Routledge, London.

Frankenberg, R. and Mani, L. 1993, 'Crosscurrents, crosstalk: Race, "postcoloniality" and the politics of location', *Cultural Studies*, vol. 7, pp. 292–310.

Frankfort-Nachmaias, C. and Nachmaias, D. 1992, *Research Methods in the Social Sciences*, 4th edn, Edward Arnold, London.

Freeland, G. 1995, *Canberra Cosmos: The Pilgrim's Guidebook to Sacred Sites and Symbols of Australia's Capital*, Primavera, Sydney.

Gahan, C. and Hannibal, M. 1998, *Doing Qualitative Research Using QSR NUD.IST*, Sage, London.

Gale, S. J. 1996, *75 years: The Anniversary of University Geography in Australia*, Department of Geography, University of Sydney, Sydney.

Game, A. 1988, 'Canberra and nation', in P. Grundy (ed.), *Canberra: A People's Capital*, Australian Institute of Urban Studies, Canberra.

Gardner, R., Neville, H. and Snell, J. 1983, Vietnamese Settlement in Springvale, Monash University Graduate School of Environmental Science, Environmental Report No.14, Melbourne.

Geertz, C. 1973a, *The Interpretation of Culture: selected essays*, Basic Books, New York.

— 1973b, 'Thick description: toward an interpretive theory of culture', in *The Interpretation of Cultures*, Basic Books, New York.

Gibson, K., Cameron, J. and Veno, A. 1999, *Negotiating Restructuring and Sustainability: A Study of Communities Experiencing Rapid Social Change*, Australian Housing and Urban Research Institute Working Paper, AHURI, Melbourne. Also available at <http://www.ahuri.edu.au.>

Gibson-Graham, J. K. 1994, '"Stuffed if I know!": Reflections on post-modern feminist social research', *Gender, Place and Culture*, vol. 1, no. 2, pp. 205–24.

Gilbert, M. 1994, 'The politics of location: doing feminist research at "home"', *Professional Geographer*, vol. 46, pp. 90–96.

Gold, R. L. 1958, 'Roles in sociological field observation', *Social Forces*, vol. 36, pp. 219–25.

Golledge, R. 1997, 'On reassembling one's life: overcoming disability in the academic environment', *Environment and Planning D: Society and Space*, vol. 15, no. 4, pp. 391–409.

Gould, P. 1988, 'Expose yourself to geographic research', in J. Eyles (ed.), *Research in Human Geography*, Blackwell, Oxford.

Goss, J. 1988, 'The built environment and social theory: towards an architectural geography' *Professional Geographer*, vol. 40, no. 4, pp. 392–403.

Goss, J. D. and Leinbach, T. R. 1996, 'Focus groups as alternative research practice: Experience with transmigrants in Indonesia', *Area*, vol. 28, no. 2, pp. 115–23.

Gregory, D. 1994, 'Paradigm', in R. J. Johnston, D. Gregory and D. M. Smith (eds), *The Dictionary of Human Geography*, Blackwell, Oxford.

— 1978, *Ideology, Science and Human Geography*, Hutchinson, London.

Gregory, S. 1998, Consuming the Cool: Children's Popular Consumption Culture, MA thesis, Department of Geography, The University of Auckland.

Griffith, D., Desloges, J. and Amrhein, C. 1990, *Statistical Analysis for Geographers*, Prentice Hall Engineering, Science and Mathematics, New Jersey.

Guelke, L. 1978, 'Geography and logical positivism', in D. Herbert and R. J. Johnston (eds), *Geography and the Urban Environment*, vol. 1, Wiley, New York.

Gupta, A. and Ferguson, J. 1997, *Anthropological Locations: Boundaries and Grounds of a Field Science*, University of California Press, Berkeley.

Hammersley, M. 1992, 'Deconstructing the qualitative-quantitative divide', in J. Brannen (ed.), *Mixing Qualitative and Quantitative Research*, Aldershot Avebury, USA.

Hammersley, M. and Atkinson, P. 1983, *Ethnography: Principles in Practice*, Tavistock, London.

Haraway, D. 1991, 'Situated knowledges. The science question in feminism and the privilege of partial perspective', in D. Haraway (ed.), *Simians, Cyborgs and Women: The reinvention of nature*, Routledge, London.

Harley, J. B. 1992, 'Deconstructing the map', in T. Barnes and J. Duncan (eds), *Writing Worlds: Discourse, Text and Metaphor in the Representation of Landscape*, Routledge, London.

Harris, C. 1997, *The Resettlement of British Columbia: Essays on colonialism and geographical change*, University of British Columbia Press, Vancouver.

Harrison, C. M. and Burgess, J. 1994, 'Social constructions of nature: a case study of conflicts over the development of Rainham marshes', *Transactions of the Institute of British Geographers*, vol. 19, no. 3, pp. 291–310.

Hartig, K. V. and Dunn, K. M. 1998, 'Roadside memorials: interpreting new deathscapes in Newcastle, New South Wales', *Australian Geographical Studies*, vol. 36, no. 1, pp. 5–20.

Hay, I., 1998, 'Making moral imaginations: research ethics, pedagogy, and professional human geography', *Ethics, Place and Environment*, vol. 1, no. 1, pp. 55–75.

— 1996, *Communicating in Geography and the Environmental Sciences*, Oxford University Press, Melbourne.

Hay, I., Bochner, D. and Dungey, C. 1997, *Making the Grade. A guide to successful communication and study*, Oxford University Press, Melbourne.

Heathcote, R. L. 1975, *Australia*, Longman, London.

Heidegger, M. 1996 (1927), *Being and Time*, translated by J. Stambaugh, State University of New York Press, Albany.

Herman, R. D. K. 1999, 'The Aloha State: place names and the anti-conquest of Hawai'i', *Annals of the Association of American Geographers*, vol. 89, no. 1, pp. 76–102.

Herod A. 1993, 'Gender issues in the use of interviewing as a research method', *Professional Geographer*, vol. 45, no. 3, pp. 305–17

Hinchliffe, S., Crang, M., Reimer, S. and Hudson, A. 1997, 'Software for Qualitative Research: 2. Some Thoughts on "Aiding" Analysis', *Environment and Planning* A, vol. 29, no. 5, pp. 1109–24.

Hirst, P. A. 1979, *On Law and Ideology*, Macmillan, London.

Hodge, S. and Costello, L. 1998, 'Sexuality and space', in E. Stratford (ed.), *Australian Cultural Geographies*, Oxford, Melbourne.

Holbrook, B. and Jackson, P. 1996, 'Shopping around: Focus group research in North London', *Area*, vol. 28, no. 2, pp. 136–42.

Holland, P. 1991, 'Poetry and landscape in New Zealand', *New Zealand Journal of Geography*, vol. 92, pp. 8–9.

Holland, P. G. and Hargreaves, R. P. 1991, 'The trivial round, the common task: work and leisure on a Canterbury Hill Country run in the 1860s and 1870s', *New Zealand Geographer*, vol. 47, no. 1, pp. 19–25.

Holland, P., Pawson, E. and Shatford, T. 1991, 'Qualitative resources in geography' [Special Issue], *New Zealand Journal of Geography*, vol. 92.

Holt-Jensen, A. 1988, 'Multiple meanings: shopping and the cultural politics of identity', *Environment and Planning* A, vol. 27, pp. 1913–30.

Hoppe, M. J., Wells, E. A., Morrison, D. M. and Wilsdon, A. 1995, 'Using focus groups to discuss sensitive topics with children', *Evaluation Review*, vol. 19, no. 1, pp. 102–14.

Huggan, G. 1995, 'Decolonizing the map', in B. Ashcroft, G. Griffiths and H. Tiffin (eds), *The Post-Colonial Studies Reader*, Routledge, London.

Jackson, J. 1990, "I am a field note": Fieldnotes as a symbol of professional identity', in R. Sanjek (ed.), *Fieldnotes: The Makings of Anthropology*, Cornell University Press, Ithica.

Jackson, P. 1992, 'Constructions of culture, representations of race: Edward Curtis's "way of seeing"', in K. Anderson and F. Gale (eds), *Inventing Places: Studies in Cultural Geography*, Longman Cheshire, Melbourne.

— 1983, 'Principles and problems of participant observation', *Geografiska Annaler*, vol. 65B, pp. 39–46.

Jackson, P. and Holbrook, B. 1995, 'Multiple meanings: shopping and the cultural politics of identity', *Environment and Planning* A, vol. 27, pp. 1913–30.

Jackson, P. A. 1993, 'Changing ourselves: a geography of position', in R. J. Johnston (ed.), *The Challenge for Geography*, Basil Blackwell, Oxford.

Jackson-Lears, T. 1985, 'The concept of cultural hegemony: problems and possibilities', *American Historical Review*, vol. 90, pp. 567–93.

Jacobs, J. M. 1999, 'The labours of cultural geography', in E. Stratford (ed.), *Australian Cultural Geographies*, Oxford University Press, Melbourne.

— 1996, *Edge of Empire: Postcolonialism and the City*, Routledge, London.

— 1993, '"Shake 'im this country": the mapping of the Aboriginal sacred in Australia—the case of Coronation Hill', in P. Jackson and J. Penrose (eds), *Constructions of Race, Place and Nation*, University College, London.

— 1992, 'Culture of the past and urban transformation: The Spitalfield Market redevelopment in East London', in K. J. Anderson and F. Gale (eds), *Inventing Places: Studies in Cultural Geography*, Longman Cheshire, Melbourne.

Jarrett, R. L. 1994, 'Living poor: Family life among single parent African American women', *Social Problems*, vol. 41, no. 1, pp. 30–49.

Jay, N. 1981, 'Gender and dichotomy', *Feminist Studies*, vol. 7, pp. 38–56.

Jick, T. D. 1979, 'Mixing qualitative and quantitative methods: triangulation in action', *Administrative Science Quarterly*, vol. 24, December, pp. 602–11.

Johnson, A. 1996, '"It's good to talk": The focus group and the sociological imagination', *The Sociological Review*, vol. 44, no. 3, pp. 517–38.

Johnson, R. B. 1997, 'Examining the validity structure of qualitative research', *Education*, vol. 118, no. 2, pp. 282–90.

Johnston, R. J. 1983, *Philosophy and Human Geography: An Introduction to Contemporary Approaches*, Edward Arnold, London.

— 1979, *Geography and Geographers: Anglo-American Human Geography Since 1945*, Edward Arnold, London.

— 1978, *Multivariate Statistical Analysis in Geography: A Primer on the General Linear Model*, Longman, New York.

Jones, A. 1992, 'Writing Feminist Educational Research: Am "I" in the Text?', S. Middleton and A. Jones (eds), in *Women and Education in Aotearoa*, Bridget Williams Books, Wellington.

Judd, C. M., Smith, E. R. and Kidder, L. H. 1991, *Research Methods in Social Relations*, 6th edn, Holt, Rinehart and Winston, Sydney.

Kearns, R. 1997, 'Constructing (Bi)cultural geographies: research on, and with, people of the Hokianga District', *New Zealand Geographer*, vol. 52, pp. 3–8.

— 1991a, 'Talking and listening: avenues to geographical understanding', *New Zealand Journal of Geography*, vol. 92, pp. 2–3.

— 1991b, 'The place of health in the health of place: the case of the Hokianga special medical area', *Social Science and Medicine*, vol. 33, pp. 519–30.

— 1987, In the Shadow of Illness: A Social Geography of the Chronically Mentally Disabled in Hamilton, Ontario, PhD dissertation, Department of Geography, McMaster University.

Kearns, R. A., Smith, C. J. and Abbott, M. W. 1991, 'Another day in paradise? Life on the margins in urban New Zealand', *Social Science and Medicine*, vol. 33, pp. 369–79.

Keen, J. and Packwood, T. 1995, 'Case study evaluation', *British Medical Journal*, vol. 311, no. 7002, pp. 444–8.

Kelle, U. 1997a, 'Capabilities for theory building and hypothesis testing in software for computer aided qualitative data analysis', *The Data Archive Bulletin*, vol. 65, no. 10, <http://dawww.essex.ac.uk/about/bulletin.html> (8 February 1999).

— 1997b, 'Theory building in qualitative research and computer programs for the management of textual data', *Sociological Research Online*, vol. 2, no. 2, <http://www.socresonline.org.uk/socresonline/2/2/1.html> (8 February 1999).

Kellehear, A. 1993, *The Unobtrusive Researcher: A Guide to Methods*, Allen and Unwin, Sydney.

Kidder, L. H., Judd, C. M. and Smith, E. R. 1986, *Research Methods in Social Relations*, CBS Publishing, New York.

Kirby, S. and Hay, I. 1997, '(Hetero)sexing space: gay men and "straight" space in Adelaide, South Australia', *Professional Geographer*, vol. 49, no. 3, pp. 295–305.

Kirk, J. and Miller, M. 1986, *Reliability and Credibility in Qualitative Research*, Sage, Beverly Hills.

Kitchin, R. and Tate, N. 2000, *Conducting Research in Human Geography: theory, methodology and practice*, Longman, London.

Kitzinger, J. 1994, 'The methodology of focus groups: The importance of interaction between research participants', *Sociology of Health and Illness*, vol. 16, no. 1, pp. 103–21.

Kivell, L. 1995, Sex-gender and race: constructing a harvest workforce, MA thesis, Department of Geography, The University of Auckland.

Kluckhohn, F. R. 1940, 'The participant observer technique in small communities', *American Journal of Sociology*, vol. 46, pp. 331–43.

Kneale, P. 1999, *Study Skills for Geography Students*, Arnold, London.

Kolakowski, L. 1972, *Positivist Philosophy: From Hume to the Vienna Circle*, Penguin, London.

Kong, L. 1999, 'Cemeteries and columbaria, memorials and mausoleums: narrative and interpretation in the study of deathscapes in geography', *Australian Geographical Studies*, vol. 37, no. 1, pp. 1–10.

— 1998, 'Refocussing on qualitative methods: problems and prospects for research in a specific Asian context', *Area*, vol. 30, no. 1, pp. 79–82.

Krueger, R. 1994, *Focus Groups: A Practical Guide for Applied Research*, 2nd edn, Sage, Thousand Oaks.

Kunstler, H. 1994, *The Geography of Nowhere*, Touchstone, USA.

Labovitz, S. and Hagedorn, R. 1981, *Introduction to Social Research*, 3rd edn, McGraw-Hill, Sydney.

Lane, R. 1997, 'Oral histories and scientific knowledge in understanding environmental change: a case study in the Tumut region, NSW', *Australian Geographical Studies*, vol. 35, no. 2, pp. 195–205.

Latimer, B. 1998, Masculinity, Place and Sport: Rugby Union and the Articulation of the 'New Man' in Aotearoa/New Zealand, MA thesis, Department of Geography, The University of Auckland.

Lawson, V. 1995, 'The politics of difference: examining the quantitative/qualitative dualism in post-structuralist feminist research', *Professional Geographer*, vol. 47, no. 4, pp. 449–57.

Le Doeff, M. 1987, 'Women and philosophy', in T. Moi (ed.), *French Feminist Thought: A reader*, Basil Blackwell, Oxford.

Lee, R. 1992, 'Teaching qualitative geography: a JGHE written symposium', *Journal of Geography in Higher Education*, vol. 16, no. 2, pp. 123–6.

Lewis, P. F. 1979, 'Axioms for reading the landscape, some guides to the American scene', in D.W. Meinig (ed.), *The Interpretation of Ordinary Landscapes: Geographical Essays*, Oxford University Press, New York.

Ley, D. 1974, *The Black Inner City as Frontier Outpost*, Monograph No. 7, Association of American Geographers, Washington, D.C.

Lincoln, Y. and Guba, E. 1985, *Naturalistic Inquiry*, Sage, Beverly Hills.

Lindsay, J. M. 1997, *Techniques in Human Geography*, Routledge, London.

Linstead, S. and Grafton-Small, R. 1990, 'Organisational bricolage', in B. Turner (ed.), *Organisational Symbolism*, Walter de Gruyter, Berlin.

Lloyd, G. 1984, *The Man of Reason: 'male' and 'female' in Western philosophy*, Methuen, London.

Longhurst, R. 1996, 'Refocusing groups: Pregnant women's geographical experiences of Hamilton, New Zealand/Aotearoa', *Area*, vol. 28, no. 2, pp. 143–49.

— 1995, 'The geography closest in—the body... the politics of pregnability', *Australian Geographical Studies*, vol. 33, pp. 214–23.

Lowenthal, D. and Prince, H. 1965, 'English landscape tastes', *Geographical Review*, vol. 47, no. 4, pp. 449–57.

Lunt, P. and Livingstone, S. 1996, 'Rethinking the focus group in media and communications research', *Journal of Communication*, vol. 46, no. 2, pp. 79–98.

Lutz, C. A. and Collins J. L. 1993, *Reading National Geographic*, University of Chicago Press, Chicago.

Maccoby, E. and Maccoby, N. 1954, 'The interview: a tool of social science', in G. Lindzey (ed.), *Handbook of Social Psychology*, Addison-Wesley, Cambridge, Massachusetts.

Mackenzie, S. 1989, *Visible Histories: Women and Environments In a Post-War British City*, McGill-Queen's University Press, Montreal.

Manning, K. 1997, 'Authenticity in constructivist inquiry: methodological considerations without prescription', *Qualitative Inquiry*, vol. 3, no. 1, pp. 93–104.

Marcus, G. E. and Fisher, M. M. J. 1986, *Anthropology as Cultural Critique: An Experimental Moment In the Human Sciences*, University of Chicago Press, Chicago.

Martin, H.P. and Schumann, H. 1997, *The Global Trap: Globalization and the assault on prosperity and democracy*, Zed, London.

Massey, D. B. 1993, 'Power-geometry and a progressive sense of place', in J. Bird, B. Curtis, G. Robertson and L. Tickner (eds), *Mapping the futures*, Routledge, London.

Massey, D. and Meegan, R. (eds) 1985, *Politics and Method—Contrasting Studies in Industrial Geography*, Methuen, London.

May, T. 1997, *Social Research: Issues, Methods and Process*, 2nd edn, Open University Press, Buckingham.

— 1993, *Social Research: Issues, Methods and Process*, Open University Press, Buckingham.

Mays, N. and Pope, C. 1997, 'Rigour and qualitative research', *British Medical Journal*, vol. 310, no. 6997, pp. 109–113.

McBride, B. 1999, 'The (post)colonial landscape of Cathedral Square: urban redevelopment and representation in the "Cathedral City"', *New Zealand Geographer*, vol. 55, no. 1, pp. 3–11.

McClean, R., Berg, L. D. and Roche, M. M. 1997, 'Responsible Geographies: Co-creating Knowledges in Aotearoa', *New Zealand Geographer*, vol. 53, no. 2, pp. 9–15.

McCracken, J. 1991, 'Looking at photographs', *New Zealand Journal of Geography*, vol. 92, pp. 12–14.

McDowell, L. 1998, 'Illusions of power: interviewing local elites', *Environment and Planning A*, vol. 30, pp. 2121–32.

— 1992, 'Valid games? A response to Erica Schoenberger', *Professional Geographer*, vol. 44, no. 2, pp. 212–15.

McDowell, L. and Court, G. 1994, 'Performing work: bodily representations in merchant banks', *Environment and Planning D: Society and Space*, vol. 12, pp. 727–50.

McKendrick, J. H. 1996, *Multi-method Research in Population Geography: A Primer to Debate*, The University of Manchester, Population Geography Research Group, Manchester.

McNeill, D. 1998, 'Writing the new Barcelona', in T. Hall and P. Hubbard (eds), *The Entrepreneurial City: Geographies of Politics, Regime and Representation*, John Wiley, London.

Mee, K. J. 1994, 'Dressing up the suburbs: representations of Western Sydney', in K. Gibson and S. Watson (eds), *Metropolis Now: Planning and the Urban in Contemporary Australia*, Pluto Press, Sydney.

Merton, R. K. 1987, 'The focussed interview and focus groups: Continuities and discontinuities', *Public Opinion Quarterly*, vol. 51, no. 4, pp. 550–66.

Miles, M. B. and Huberman, A. M. 1984, *Qualitative Data Analysis: a sourcebook of new methods*, Sage, Beverly Hills.

Minh-Ha, Trinh T. 1991, *When the Moon Waxes Red: Representation, Gender and Cultural Politics*, Routledge, New York.

Minichiello, V., Aroni, R., Timewell, E. and Alexander, L. 1995, *In-Depth Interviewing: Principles, Techniques, Analysis*, 2nd edn, Longman Cheshire, Melbourne.

— 1990, *In-Depth Interviewing: Researching People*, Longman Cheshire, Melbourne.

Mishler, E. 1986, *Research Interviewing: Context and Narrative*, Harvard University Press, Cambridge, Massachusetts.

Mitchell, W. J. T. 1986, *Iconology: Image, Text, Ideology*, University of Chicago Press, Chicago.

Mohanty, C. T. 1991, 'Cartographies of struggle: Third world women and the politics of feminism', in C. Mohanty, A. Russo, and L. Torres (eds), *Third World Women and the Politics of Feminism*, University of Indiana Press, Bloomington.

Monk, J. and Hanson, S. 1982, 'On not excluding half of the human in human geography', *Professional Geographer*, vol. 34, pp. 11–23.

Morgan, D. L. 1997, *Focus Groups as Qualitative Research*, 2nd edn, Sage, Thousand Oaks.

Morgan, D. L. 1996, 'Focus groups', *Annual Review of Sociology*, vol. 22, pp. 129–52.

Morgan, D. L. (ed.) 1993, *Successful Focus Groups: Advancing the State of the Art*, Sage, Newbury Park.

Morgan, D. L. and Krueger, R. A. 1993, 'When to use focus groups and why', in D. L. Morgan (ed.), *Successful Focus Groups: Advancing the State of the Art*, Sage, Newbury Park.

Moser, C. A. and Kalton, G. 1983, *Survey Methods in Social Investigation*, Heinemann, London.

Moss, P. 1995, 'Reflections on the "gap" as part of the politics of research design', *Antipode*, vol. 27, no. 1, pp. 82–90.

Mostyn, B. 1985, 'The content analysis of qualitative research data: a dynamic approach', in M. Brown, J. Brown and D. Canter (eds), *The Research Interview: Uses and Approaches*, Academic Press, London.

Muller, S. 1999, Myths, media and politics. Implications for koala management decisions in Kangaroo Island, South Australia, paper presented to Institute of Australian Geographers' conference, 27 September to 1 October, Sydney.

Myers, G. 1998, 'Displaying opinions: Topics and disagreement in focus groups', *Language in Society*, vol. 27, pp. 85–111.

Myers, G., Klak, T. and Koehl, T. 1996, 'The inscription of difference: news coverage of the conflicts in Rwanda and Bosnia', *Political Geography*, vol. 15, no. 1, pp. 21–46.

Myers, G. and Macnaghten, P. 1998, 'Rhetorics of environmental sustainability: Commonplaces and places', *Environment and Planning* A, vol. 30, pp. 333–53.

Nast, H. 1994, 'Opening remarks: Women in the field', *Professional Geographer*, vol. 46, pp. 54–66.

Nelson, C., Treichler, P. and Grossberg, L. 1992, 'Cultural studies: an introduction', in L. Grossberg, C. Nelson and P. Treichler (eds), *Cultural Studies*, Routledge, New York.

Nettlefold, P. A. and Stratford, E. 1999, 'The production of climbing landscapes-as-texts', *Australian Geographical Studies*, vol. 37, no. 2, pp. 130–41.

Nietzsche, F. W. 1969, *On the Genealogy of Morals*, translated by W. Kaufmann and R. J. Hollingdale, Vintage, New York.

Oakley, A. 1981, *From Here to Maternity: Becoming a Mother*, Penguin, Harmondsworth.

O'Brien, K. 1993, 'Improving survey questionnaires through focus groups', in D. L. Morgan (ed.), *Successful Focus Groups: Advancing the State of the Art*, Sage, Newbury Park.

O'Donnell, D. J. and Layder, D. 1994, *Methods, Sex and Madness*, Routledge, London.

Opie, A. 1992, 'Qualitative Research, appropriation of the other and empowerment', *Feminist Review*, vol. 40, pp. 52–69.

Oritz, S. M. 1994, 'Shopping for sociability in the mall', *Research in Community Sociology*, vol. 4 (supplement), pp. 183–99.

Parr, H. 1998, 'Mental health, ethnography and the body', *Area*, vol. 30, pp. 28–37.

Patton, M. Q. 1990, *Qualitative Evaluation and Research Methods*, 2nd edn, Sage, Beverly Hills.

Pawson, E. 1991, 'Monuments, memorials and cemeteries: icons in the landscape', *New Zealand Journal of Geography*, vol. 92, pp. 26–7.

Peace, R. 1998, 'CAQDAS/NUD*IST: computer assisted qualitative data analysis software/non-numerical, unstructured data. Indexing, searching and theorising – a geographical perspective', in E. Bliss (ed.), *Proceedings of the Second Joint Conference, Institute of Australian Geographers and New Zealand Geographical Society*, January 1997, New Zealand Geographical Society, Hamilton, pp. 382–85.

Pearson, L. J. 1996, Place Re-identification: the 'Leisure Coast' as a Partial Representation of Wollongong, BSc Honours thesis, School of Geography, University of New South Wales.

Pearson, L. J. and Dunn, K. M. 1999, 'Reidentifying Wollongong: dispossession of the local citizenry', *Proceedings of the Australian University Tourism and Hospitality Education 1999 National Research Conference*, Adelaide.

Philip, L. J. 1998, 'Combining quantitative and qualitative approaches to social research in human geography—an impossible mixture?', *Environment and Planning A*, vol. 30, pp. 261–76.

Pickles, J. 1992, 'Texts, hermeneutics and propaganda maps', in T. J. Barnes and J. S. Duncan (eds), *Writing Worlds*, Routledge, London.

Pile, S. 1992, 'Oral history and teaching qualitative methods', *Journal of Geography in Higher Education*, vol. 16, no. 2, pp. 135–43.

Poland, B. D. 1995, 'Transcription quality as an aspect of rigor in qualitative research', *Qualitative Inquiry*, vol. 1, no. 3, pp. 290–310.

Ponga, M. 1998, 'I Nga ra o Mua: (Re)Constructions of symbolic layers in Te Poho-o-Hinemihi Marae', MA thesis, Department of Geography, The University of Auckland.

Popper, K. 1959, *The Logic of Scientific Discovery*, Hutchinson, London.

Porteous, D. J. 1985, 'Smellscape', *Progress in Human Geography*, vol. 9, pp. 356–78.

Powell, J. M. 1988, *An Historical Geography of Modern Australia*, Cambridge University Press, Cambridge.

Pulvirenti, M. 1997, 'Unwrapping the parcel: an examination of culture through Italian home ownership', *Australian Geographical Studies*, vol. 35, no. 1, pp. 32–9.

Reason, P. and Rowan, J. (eds) 1981, *Human Inquiry: A sourcebook of new paradigm research*, John Wiley and Sons, Chichester.

Reed, M. and Harvey, D. 1992, 'The new science and the old: complexity and realism in the social sciences', *Journal for the Theory of Social Behaviour*, vol. 22, no. 4, pp. 353–80.

Richards, L. 1997, 'Computers and Qualitative Analysis', *The International Encyclopedia of Education*, Elsevier Science, Oxford.

— 1990, *Nobody's Home: Dreams and Realities in a New Suburb*, Oxford University Press, Melbourne.

Richards, L. and Richards, T. 1995, 'Using hierarchical categories in qualitative data analysis', in U. Kelle (ed.), *Computer-Aided Qualitative Data Analysis: Theory, Methods and Practice*, Sage, London. Also available at <http://www.qsr.com.au/otherinfo/papers/hierarchies.html> (8 February 1999).

Richardson, L. 1994, 'Writing. A Method of Inquiry', in N. K. Denzin and Y. S. Lincoln (eds), *Handbook of Qualitative Research*, Sage, Thousand Oaks.

Rimmer, P. J. and Davenport, S. 1998, 'The geographer as itinerant: Peter Scott in flight, 1952–1996', *Australian Geographical Studies*, vol. 36, no. 2, pp. 123–42.

Roberts, J. and Sainty, G. 1996, *Listening to the Lachlan*, Sainty and Associates, Potts Point.

Robertson, B. M. 1994, *Oral History Handbook*, Oral History Association of Australia, Adelaide.

Robinson, G. 1998, *Methods and Techniques in Human Geography*, John Wiley and Sons, Chichester.

Roche, M. 1991, 'A picture is worth a thousand words: cartoons and social commentary', *New Zealand Journal of Geography*, vol. 92, pp. 10–11.

Rodaway, P. 1994, *Sensuous Geographies: Body, Sense, Place*, Routledge, London.

Rose, C. 1988, 'The concept of reach and the Anglophone minority in Quebec', in J. Eyles and D. Smith (eds), *Qualitative Methods in Human Geography*, Polity Press, Cambridge.

Rose, G. 1997, 'Situating knowledges: Positionality, reflexivities and other tactics', *Progress in Human Geography*, vol. 21, pp. 305–20.

Rose, G. 1993, *Feminism and Geography*, University of Minnesota Press, Minneapolis.

Rothenberg, T. Y. 1993, 'Voyeurs of imperialism: The National Geographic Magazine before World War II', in A. Godlewska and N. Smith (eds), *Geography and Empire*, Blackwell, Oxford.

Rowles, G. D. 1978, *Prisoners of Space: Exploring the Geographical Experience of Older People*, Westview Press, Boulder.

Sanjek, R. (ed.) 1990, *Fieldnotes: the Makings of Anthropology*, Cornell University Press, Ithaca.

Sarantakos, S. 1993, *Social Research*, MacMillan, South Melbourne.

Sausseur, F. de, 1983, *Course in General Linguistics*, Duckworth, London.

Sayer, A. 1992, *Method in Social Science*, 2nd edn, Routledge, London.

Sayer, A. and Morgan, K. 1985, 'A modern industry in a declining region: links between method, theory and policy', in D. Massey and R. Meegan (eds), *Politics and Method: Contrasting Studies in Industrial Geography*, Methuen, London.

Schein, R. H. 1997, 'The place of landscape: a conceptual framework for interpreting an American scene', *Annals of the Association of American Geographers*, vol. 87, no. 4, pp. 660–80.

Schoenberger, E. 1992, 'Self-Criticism and Self-Awareness in Research: A Reply to Linda McDowell', *Professional Geographer*, vol. 44, pp. 215–18.

— 1991, 'The corporate interview as a research method in economic geography', *Professional Geographer*, vol. 43, no. 2, pp. 180–89.

School of Global Studies, 1998, *Essay Format and Essay Writing for Geography Students*, School of Global Studies, Massey University, Palmerston North.

Scott, K., Park, J., Cocklin, C. and Blunden, G. 1997, *A Sense of Community: An Ethnography of Rural Sustainability in the Mangakahia Valley, Northland*, Occasional Publication 33, Department of Geography, The University of Auckland.

Seebohm, K. 1994, 'The nature and meaning of the Sydney Mardi Gras in a landscape of inscribed social relations', in R. Aldrich (ed.), *Gay Perspectives II: More Essays in Australian Gay Culture*, Department of Economic History with the Australian Centre for Gay and Lesbian Research, University of Sydney, Sydney.

Sekaran, U., 1992, *Research Methods for Business: A Skill Building Approach*, 2nd edn, John Wiley and Sons Inc., New York.

Shapcott, M. and Steadman, P. 1978, 'Rhythms of urban activity', in T. Carlstein, D. Parkes and N. Thrift (eds), *Human Activity and Time Geography*, John Wiley and Sons, New York.

Shaw, G. and Wheeler, D. 1994, *Statistical Techniques in Geographical Analysis*, Halsted Press, New York.

Sherraden, M. n.d., *How to do Focus Groups* [on line], available at <http://www.gwbssw.wustl.edu/fgroups/fghowto.html>.

Silverman, D. 1993, *Interpreting Qualitative Data: Methods for Analysing Talk, Text and Interaction*, Sage, London.

Smith, S. J. 1994, 'Soundscape', *Area*, vol. 26, pp. 232–40.

— 1988, 'Constructing local knowledge: The analysis of self in everyday life', in J. Eyles and D. M. Smith (eds), *Qualitative Methods in Human Geography*, Polity Press, Cambridge.

— 1981, 'Humanistic method in contemporary social geography', *Area*, vol. 15, pp. 355–58.

Sparke, M. 1998, 'A map that roared and an original atlas: Canada, cartography and the narration of a nation', *Annals of the Association of American Geographers*, vol. 88, no. 3, pp. 463–95.

Spate, O. H. K. and Learmonth, A. 1967, *India and Pakistan: A General and Regional Geography*, Methuen, London.

Spradley, J. P. 1980, *Participant Observation*, Holt, Rinehart and Wilson, New York.

Srivinas, M. N., Shah, A. M. and Ramaswamy, E. A. 1979, *The Fieldworker and the Field: Problems and Challenges in Sociological Investigation*, Oxford University Press, Delhi.

Stake, R. 1995, *The Art of Case Study Research*, Sage, Thousand Oaks.

Stanton, N. 1996, *Mastering Communication*, 3rd edn, MacMillan, London.

Stewart, D. W. and Shamdasani, P. N. 1990, *Focus Groups: Theory and Practice*, Sage, Newbury Park.

Stratford, E. (ed.) 1999, *Australian Cultural Geographies*, Oxford University Press, Melbourne.

— 1998, 'Public spaces, urban youth and local government: the skateboard culture in Hobart's Franklin Square', in R. Freestone (ed.), *20th Century Urban Planning Experience, Proceedings of the 8th International Planning History Conference*, University of New South Wales, Sydney.

Strauss, A. and Corban, J. 1990, *Basics of Qualitative Research. Grounded Theory, Procedures and Techniques*, Sage, Newbury Park.

Sudman, S. and Bradburn, N. M. 1982, *Asking Questions: A Practical Guide to Questionnaire Design*, Jossey-Bass, San Francisco.

Swenson, J. D., Griswold, W. F. and Kleiber, P. B. 1992, 'Focus groups: Method of inquiry/intervention', *Small Group Research*, vol. 23, no. 4, pp. 459–74.

Symposium on Computing and Qualitative Geography, 1995, University of Durham, 11–12 July, <http://www.helsinki.fi/neu/lists/qual-software/0034.html> (8 February 1999).

Tesch, R. 1989, 'Computer Software and Qualitative Analysis: A Reassessment', in G. Blank, E. Brent and J. L. McCartney (eds), *New Technology in Sociology: Practical Applications in Research and Work*, Transaction Books, New Brunswick.

Thompson, S. 1994, 'Suburbs of opportunity: the power of home for migrant women' in K. Gibson and S. Watson (eds), *Metropolis Now: Planning and the Urban in Contemporary Australia*, Pluto, Sydney.

Thrift, N. J. 1996, *Spatial Formations*, Sage, Thousand Oaks.

Tremblay, M. A. 1982, 'The key informant technique: a non-ethnographic application', in R. G. Burgess (ed.), *Field Research: A Sourcebook and Field Manual*, Allen and Unwin, London.

Tuan, Y. F. 1991, 'Language and the making of place: a narrative-descriptive approach', *Annals of the Association of American Geographers*, vol. 81, no. 4, pp. 684–96.

Valentine, G. 1993, '(Hetero)sexing space: lesbian perceptions and experiences of everyday spaces', *Environment and Planning D: Society and Space*, vol. 11, no. 4, pp. 395–413.

— 1989, 'The geography of women's fear', *Area*, vol. 21, no. 4, pp. 385–90.

Waitt, G. and McGuirk, P. M. 1996, 'Marking time: tourism and heritage representation at Millers Point, Sydney', *Australian Geographer*, vol. 27, no. 1, pp. 11–29.

Walmsley, D. S. and Lewis G. I. 1984, *Human Geography: Behavioural Approaches*, Longman, New York

Walsh, B. and Lavalli, T. 1996, 'Comparative Review of NUD*IST, Atlas/ti, Folio Views', *Microtimes*, no.162, <http://www.microtimes.com/162/research.html> (8 February 1999).

Ward, B., 1972, *What's Wrong with Economics?*, MacMillan, London.

Ward, V. M., Bertrand J. T. and Brown, L. F. 1991, 'The comparability of focus group and survey results: Three case studies', *Evaluation Review*, vol. 15, no. 2, pp. 266–83.

Warde, A. 1989, 'Recipes for a pudding: a comment on locality', *Antipode*, vol. 21, no. 3, pp. 274–81.

Watson, S. 1991, 'Gilding the smokestacks: the new symbolic representations of deindustrialised regions', *Environment and Planning D: Society and Space*, vol. 9, no. 1, pp. 59–70.

Wearing, B. 1984, *The Ideology of Motherhood: A Study of Sydney Suburban Mothers*, George Allen and Unwin, Sydney.

Webb, B. 1982, 'The art of note-taking', in R. G. Burgess (ed.), *Field Research: A Sourcebook and Field Manual*, Allen and Unwin, London.

Webb, E. J., Campbell, D. T., Schwartz, R. D. and Sechrest, L. 1966, *Unobtrusive Measures: Non-reactive Research in the Social Sciences*, Rand McNally, Chicago.

Weinstein, D. and Weinstein, M. 1991, 'Georg Simmel: sociological ~~flaneur~~ bricoleur', *Theory, Culture and Society*, vol. 8, pp. 151–68.

Weitzman, E. and Miles, M. 1995, *Computer Programs for Qualitative Data Analysis: Software Sourcebook*, Sage, Thousand Oaks.

Welsh, I. 1996, *Trainspotting*, W. W. Norton, London.

Western, J. C. 1981, *Outcast Cape Town*, University of Minnesota Press, Minneapolis.

White, P. and Jackson, P. A. 1995, '(Re)theorising population geography', *International Journal of Population Geography*, vol. 1, pp. 111–23.

Whyte, W. F. 1982, 'Interviewing in field research', in R. G. Burgess (ed.), *Field Research: A Sourcebook and Field Manual*, Allen and Unwin, London.

Whyte, W. H. 1957, *Street Corner Society*, University of Chicago Press, Chicago.

Wilson, A. G. 1972, 'Theoretical Geography: Some Considerations', *Transactions of the Institute of British Geographers*, vol. 7, pp. 31–44.

Winchester, H. P. M. 2000, 'Qualitative research and its place in geography', in I. Hay (ed.), *Qualitative Research Methods in Human Geography*, Oxford University Press, Melbourne.

— 1999, 'Interviews and questionnaires as mixed methods in population geography: The case of lone fathers in Newcastle, Australia', *Professional Geographer*, vol. 51, no. 1, pp. 60–67.

— 1996, 'Ethical issues in interviewing as a research method in human geography', *Australian Geographer*, vol. 27, no. 1, pp. 117–31.

— 1992, 'The construction and deconstruction of women's role in the urban landscape' in F. Gale and K. Anderson (eds), *Inventing Places: Studies in Cultural Geography*, Longman Cheshire, Melbourne.

Winchester, H. P. M. and Costello, L. N. 1995, 'Living on the street: social organisation and gender relations of Australian street kids', *Environment and Planning D: Society and Space*, vol. 13, pp. 329–48.

Winchester, H. P. M. and Dunn, K. M. in press, 'Cultural geographies of film: tales of urban reality', in K. J. Anderson and F. Gale (eds), *Inventing Places: Studies in Cultural Geography*, Addison Wesley Longman, Melbourne.

Winchester, H. P. M., Dunn, K. M. and McGuirk, P. M. 1997, 'Uncovering Carrington', in R. J. Moore and M. J. Ostwald (eds), *Hidden Newcastle: Urban Memories and Architectural Imaginaries*, Gladfly Media, Ultimo.

Winchester, H. P. M., McGuirk, P. M. and Everett, K. 1999, 'Celebration and control: Schoolies Week on the Gold Coast Queensland', in E. Teather (ed.), *Embodied Geographies: Spaces, Bodies and Rites of Passage*, Routledge, London.

Wood, L. and Williamson, S. 1996, *Consultants' Report on Franklin Square: Users, Activities and Conflicts*, UNITAS Consulting, Hobart.

Wolcott, H. 1990, 'On seeking—and rejecting—validity in qualitative research', in E. Eisner and A. Peshkin (eds), *Qualitative Inquiry In Education: the continuing debate*, Teachers College Press, New York.

Wooldridge, S. W. 1955, 'The status of geography and the role of fieldwork', *Geography*, vol. 40, pp. 73–83.

Wrigley, E. A. 1970, 'Changes in the philosophy of geography', in R. J. Chorley and P. Haggett (eds), *Frontiers in Geographical Teaching*, Methuen, London.

Zeigler, D. J., Brunn, S. D. and Johnson, J. H. 1996, 'Focusing on Hurricane Andrew through the eyes of the victims', *Area*, vol. 28, no. 2, pp. 124–9.

Index